MIKOYAN-GUREVICH

MiG-21

ALL MAJOR VARIANTS &
EXTENDED VIETNAM WAR SECTION

THEODORE GIANNA

THUNDERBOLT 47 PUBLICATIONS

Front cover photo by Rob Schleiffert

ISBN: 978-0-6456548-0-6 (Paperback)
First printing edition 2023

INTRODUCTION

The MiG-21 was the first supersonic fighter from the Soviet Union. It was first built in the mid-1950s, around the same time as the US Century Series jet fighters, the F-100, F-101, F-102, F-104, F-105, and F-106. However, the MiG 21 would outlive all those fighters.

The three largest MiG factories in the USSR manufactured the MiG-21s in many variants. The Moscow Gorki plant (no.21) saw the most extensive production; 5,278 units. The second facility, also in Moscow, was the 'Znamya Truda' (Moscow Aircraft Production Association) plant, which produced 3,203 units, and the no. 31 plant in Tbilisi, Georgia, built 1,677 units. Production lasted 27 years, from 1959 until 1986. MiG fighters for internal and export use totaled an amazing 10,158 MiG-21s built, including the last of a great breed, 2,030 MiG-21bis fighters, and 1,133 MiG-21 UM 'Mongol' training versions. It was the most mass-produced supersonic fighter aircraft in aviation history, and it still holds that title.

Although the MiG-21 was an excellent jet fighter, one aircraft had its measure, the McDonnell Douglas F-4 Phantom. The air battles that took place between the two rival jets during the long Vietnam conflict are legendary. This book devotes a section on the MiG's operations and performance in the Vietnam War.

ACKNOWLEDGMENTS

I want to thank Dale Volkov, who has the most interesting and informative Russian aviation website on the internet, for allowing me to use many photos from his website for this book. Without his assistance, this book would not be the same. His website is: https://www.airwar.ru/

Another person I wish to credit is Richard Vandervord for his outstanding MiG-21 photographs. He is one of the best aviation photographers going around. He previously provided me with many photos for an airline book that I published last year, and he has done so again for this book.

I must also compliment the USAF, US Navy, US Army, CIA, DTIC, and AFHRA for their vast archives of war records, reports, battle information, aircraft specifications, and so on from the Vietnam War, WW2, Korean War, secret wars and other conflicts that I have amassed over the years and have aided me in my research.

MIKOYAN-GUREVICH MiG-21

THE POWER OF THE 'MiG'

During the Cold War, the 'MiG' term was popular, thanks to the MiG-15, the small Soviet fighter jet that saw action in the Korean War. The phrase 'MiG' gained traction with news agencies, Western Governments, and the Western public following the Russian jet's successes in the second half of 1950 during that conflict.

The MiG-15 created concerns for the US due to its agility and swiftness. The lethal Russian plane could not be beaten by the US fighter jets. It was in late 1950, with the introduction of the Russian fighter into the Korean conflict, and instantly challenged the USAF's less agile F-80 Shooting Stars and F-84 Thunderjets, as well as the US Navy's F2H Banshees and F9F Panthers. The Americans were worried.

Whenever Americans saw a Russian aircraft on their television screens, they guessed it was one of those rotten MiGs. It didn't matter if it was a Sukhoi or a Yakovlev fighter which could also do some damage; to the audience, it had to be a MiG. That single acronym wielded enormous power during the Cold War.

It wasn't until the F-86 Sabre appeared in the Korean War in December 1950 and outperformed the MiG-15 that things began to return to normality. On the other hand, the MiG company had already created an aura of power with its jet fighter, which the term MiG represented. From the Korean War to the Vietnam War and up till the current day, Western pilots and the general public both loathed and admired the MiG fighter aircraft.

MiG was also a very popular term in the USSR. The firm was well-known within the USSR well before the Korean War and wielded political power. For example, in the late morning of April 24, 1946, when the MiG-9 jet fighter made its inaugural flight (which only lasted six minutes), it placed the MiG jet in the history books as the first Soviet jet fighter to take to the skies.

However, there was another angle to that story. The engineers of the Yakovlev firm, another renowned Soviet aeronautical design company, were discouraged by the MiG-9 making that first flight. Their design, the Yakovlev Yak-15 jet fighter, was ready to go, whereas the MiG-9 was not even at the airfield. Yet, that's how it was.

On the morning of April 24, well before the MiG-9 was set to fly, the Yakovlev firm's leader, Yakovlev himself, was not sure about the Yak-15 being ready to fly and hesitated in making a decision, which caused a slight delay. However, the engineers were not as concerned; they knew the Yak-15 was 100% ready.

Moreover, the Yak-15 was already at the airfield that morning, ready to go. However, ATC informed the Yakovlev firm that due to extensive air training for the upcoming Labor Day Air Show, all morning flights were canceled until further notice. That unnerved the engineers, who nervously waited for the training flights to end so that they could get their plane in the air. At about the same time, MiG's head, Artem Mikoyan, found out about the flight cancelations and immediately telephoned the Moscow Military District Commander to make an exception for the MiG company and allow the first flight to be made by the MiG-9, not the Yak-15. The Commander said he would do anything for Mikoyan, and so it was. Not long after that call, the MiG-9 was seen parked near the Yak-15 at the airfield.

As Yak's ground crew were preparing their plane for takeoff, they suddenly saw the MiG-9 slowly taxi to the runway threshold, and, without stopping, it made a takeoff run and flew up to the wild blue yonder. Just like that.

Thanks to the political pull that Mikoyan and the MiG firm had, the MiG-9 was the first in the air. Yak's crew were downcast, but their Yak-15 made a perfect 'second' first flight if that was any consolation.

PLANNING FOR THE FUTURE

Until the mid-1950s, the MiG Design Bureau's factories were principally dedicated to the ongoing manufacturing of the MiG-15, MiG-17, and MiG-15UTI two-seater trainers. In addition, various modifications were made by the production plants to further improve flight, tactical data, and aircraft maintenance in Air Force units. Work on prototype aircraft design and construction focused on developing new frontline fighters, fighter interceptors, and reconnaissance aircraft.

Following Moscow Aviation Institute's (MAI) order of July 8, 1952, a branch of the OKB-155 was established at Gorky, plant No 21, under the leadership of Deputy Chief Designer P. Syrova to relieve the Design Bureau of the work involved in improving the production of aircraft. (OKB was MiG's 'Bureau of Experimental Design').

The design team got to work on designing two new jets for the Soviet Air Force; the famed MiG-19 and MiG-21 fighters were the results. By the beginning of the 1960s, the MiG-19 and MiG-21 fighters were built, tested, and mass-produced. However, to get to that point, various experimental aircraft were made in small batches to test and refine the aircraft's characteristics and operability. Thus, controllability, small weapons, propulsion systems, landing aids, and ejector seats were among the serious issues addressed for the new designs. The experimental aircraft made for the tests were designated the 'E' series.

From the late 1950s to 1965, the MiG Design Bureau was also busy fulfilling the Government's assignment to build interceptors equipped with the 'Uragan-5' automatic seeker and homing system to be used on the experimental E-150, E-152, and E-152A experimental jets. During the same period, the basis was laid for designing and building the next generation of fighters like the MiG-23 and MiG-25. Together with the creation of those new types, the improvements and upgrading of the MiG-19 and MiG-21 fighters were ongoing.

THE SEARCH FOR NEW FIGHTERS

In the early to mid-1950s, the Americans started producing excellent fighter aircraft at a constant rate. It started on May 25, 1953, when the first Century series of jet fighters, the YF-100A, powered by a Pratt & Whitney XJ57-P-7 engine, made its historic flight. With the renowned pilot George Welch at the controls, the American jet broke the sonic barrier in level flight on that May day by reaching Mach 1.05. The F-100 was followed by the F-101, F-102, F-104, F-105, and the F-106. All supersonic jets, and all very capable in an air battle.

In 1953, the Soviet Union still had no fighter plane to take on the American jets. The prototype MiG-19 designated SM-2/1, had already made its first flight on May 24, 1952; however, because of testing, powerplant delays, and other requirements, the Mig-19 was almost three years away from being put into production and service.

During the development of the MiG-19, the MiG Design Bureau was also busy developing another fighter plane. The goal was to create a light supersonic frontline fighter with a tiny airframe and an excellent thrust-to-weight ratio. This would provide significant economic, operational, and tactical advantages in structural weight, per-flight fuel consumption, and the ability to land on airfields with uneven runways and taxiways. MiG's OKB-155 'Bureau of Experimental Design' was formally assigned the responsibility of building the aircraft on September 9, 1953, together with 'TsAGI' (Central Institute of

Aerodynamics), in line with a government decree.

The initial order was to construct an experimental frontline fighter with a delta-wing which would be powered by the Tumansky AM-11 turbojet engine. However, the MiG firm, like other Soviet aeronautical design companies at the time, had no experience in building delta-wing aircraft. MiG was experienced in constructing swept-wing aircraft, as the company had worked out the aerodynamics of swept-wing design via wind tunnel and full-scale flight tests. The new delta design was an unknown factor for Mig.

Nonetheless, the government request for the new fighter gave rise to two MiG brand supersonic frontline fighter aircraft, notably the E-2, also known as the MiG-23 (Faceplate) and the E-5 (Fishbed A).

The two experimental jets were built as single-engine, single-seat mid-planes with extended cigar-shaped fuselages with circular openings on the nose and rear. Both jets' wings were swept back. Thus, the only thing that distinguished one from the other was the aerodynamic design of the wings. The E-2 had sweptback wings, set at an angle of 55 degrees, while the E-5 had delta wings with a leading edge sweepback of 57 degrees.

Mikoyan labored meticulously on the new aircraft, enlisting the help of all available professionals, including theorists, engine builders, electronics technicians, and aerodynamicists. Of course, one might assemble a team of professionals to assist with designing and building a plane, but no one could ever be certain that it would be a success. For example, the project's aerodynamic specialists anticipated that around Mach 1.5, a new aerodynamic barrier lay in wait, but they didn't identify what it was since they didn't know. That meant Mikoyan was creating a design with no idea what the end result would be.

As for the aerodynamicists, the issue of the effects on aircraft at relatively high supersonic speeds was a worrying one. The supersonic speed range begins when the free stream speed exceeds Mach 1.3. (between Mach 1.3–5.0, with speeds of 915-3,806 mph). Because of fundamental variations in the behavior of flows above Mach 1, an aircraft built to fly at supersonic speeds displays significant variances in its aerodynamic design. For example, the plane must have sharp edges, small airfoil sections, and all-moving tailplanes and canards to adapt to its new environment.

The testing of the experimental jets did not begin as the designers expected. Due to technical issues, the production of the Tumansky AM-11 turbojet was delayed. To avoid wasting time while waiting for the powerplant to become available, MiG opted to begin factory flight testing of the two E-2 and E-5 prototypes with the temporary installation of the weaker 7,165 lb. (3,250 kg) AM-9 type engine (later designated RD-9).

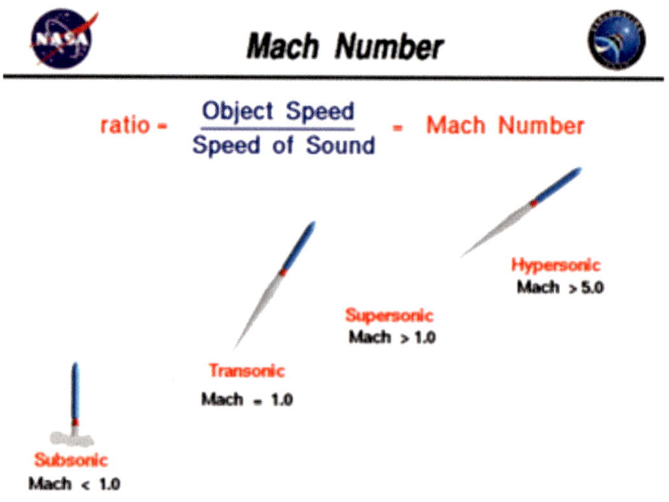

This NASA graphic shows the supersonic speed range which lies between Mach 1.3 and Mach 5. Any jet flying beyond Mach 1.3 up to Mach 5 would need heavy modifications made to its design, to be able to fly effortlessly. (Graphic by NASA, Tom Benson)

THE E-2/E2A

On December 25, 1954, the prototype E-2 jet with a leading-edge sweepback of 55 degrees was completed. But, it was powered by the weaker Tumansky AM-9 turbojet engine. The E-2 took to the skies for the first time on February 14, 1955, with the test pilot G.K. Mosolov at the controls.

Due to the low thrust of its AM-9 powerplant, the experimental E-2 (Fishbed B) did not meet the required performance during the flight tests. As a result, the E-2 was restricted to evaluating controllability, stability, and aerodynamic qualities at moderate speeds. These tests, however, were soon terminated. Consequently, the E-2 was fitted with the more powerful Tumansky AM-11 (R-37F) with 8,708 lb. (3,950 kg) military power thrust and 13,227 lb. (6,000 kg) thrust with afterburner.

The E-2 was promptly renamed E-2A, Type 63 (NATO: Faceplate), after receiving the more powerful AM-11 engine (The engine was later redesignated RD-11). Furthermore, in place of the automated slots, MiG fitted a pair of wing fences, one on each wing, of the E-2A at the midway point of the span.

When the E-2's automated slots on the leading edge of the wings were extended asymmetrically, they exhibited slightly chaotic behavior. However, the aerodynamic fences' installation helped deflect air currents away from the wing roots while boosting the effectiveness of the ailerons during maneuvers with significant angles of attack. One other addition to the E-2A was a pitot tube on the nose.

On February 17, 1956, the freshly upgraded E-2A (ex-E-2) (Faceplate) prototype took to the skies with the new Tumansky R-11-300 engine. Grigory A. Sedov was the test pilot. There were no significant issues with the flight. Then, on March 26, 1956, while flying across the Soviet Union in the same aircraft, Sedov set a speed record of 1,180 mph (1,900 kph).

OKB MiG received an order to build a prototype of the second E-2A prototype, which was designated E-2A/2 (Faceplate).

This is the E-2 experimental, built by MiG's OKB with 55 degree sweptback wings, and AM-11 engine. It made its first flight on February 14, 1955. (courtesy airwar.ru)

On July 28, 1956, the second prototype was finished, and five weeks later, on September 4, 1956, it accomplished its first flight. Grigory A. Sedov, the record-breaking test pilot, was at the controls once again. However, there were issues. Factory testing of both E-2As was beset by problems with the ineffectual RD-11 (AM-11) turbojet engine. Even the RD-11 was not strong enough for these jets. As a result, the two E-2A prototypes' flight tests had to be suspended numerous times.

The E-2A and the E-5 prototype were displayed during the Tushino Air Show near Moscow on June 24, 1955. Western observers quickly allocated the NATO designation of 'Faceplate' to the E-2A, while the delta wing E-5 became the 'Fishbed.' By this time, it had become clear to MiG that the delta wing E-5 offered much more promise for further development. In contrast, the E-2A/1 and E-2A/2 tests were often interrupted due to the engines' poor performance, which necessitated regular engine changes. On December 31, 1957, the two aircraft were submitted for state testing, and the E-2A/1 and E-2A/2 project was declared cancelled.

The prototype E-2A with the two wing fences. (courtesy airwar.ru)

The E-2A taxiing to its takeoff position, on one of its test flights. (courtesy airwar.ru)

MIKOYAN GUREVICH E-2A

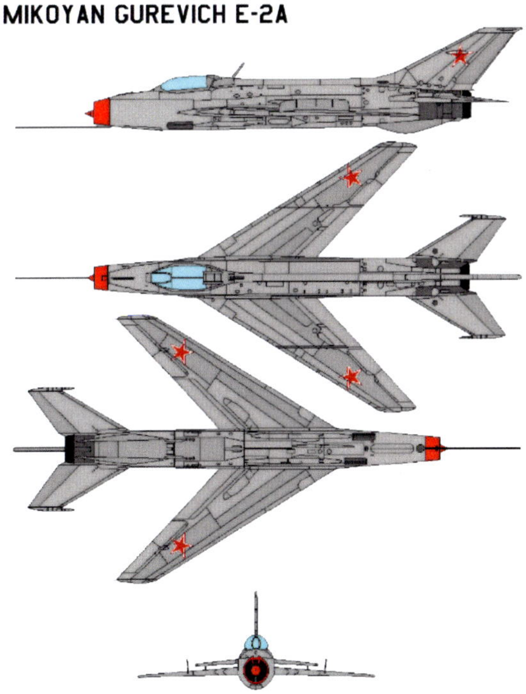

Image by bagera3005, Deviantart.com

MiG-23 - (E-2A)

The title of this section is misleading. This Mig-23 model differs from the variable-geometry fighter aircraft that entered service in 1970. The MiG design bureau, in 1957, designated this plane as the MiG-23. However, this model was only a serial modification of the E-2A and had a Tumasnky R-11-300 type engine with 11,243 lb. thrust (5,100 kg), instead of the RD-11 (AM-11) that the original E-2A had.

The MiG-23 was introduced as a small series of seven aircraft from Gorky plant no. 21. All seven MiG-23s were constructed in 1957. However, this aircraft's development was terminated in 1958, and the units produced were utilized for special test projects.

E-50/1 /2 /3

In 1955, an attempt was made to upgrade the E-2 by restructuring its rear fuselage and installing a D-S-155 rocket engine within the E-50's tail, just above the nozzle of the E-2's TRD-9E turbojet engine.

TYPE	POWERPLANT	DIMENSIONS WINGSPAN / LENGTH / WING AREA	WEIGHTS Empty / Gross / Max. t.o.w.	SPEED Max Speed	SERVICE CEILING / RANGE
E-2A Type 63	Tumansky RD-11 (AM-11) turbojet engine with 8,708 lb. thrust, 13,227 lb. with ab	26.6 ft / 43 ft. 4 in. / 226.04 sq ft	9568 lb. / 13,778 lb. / 13,779 lb.	1,181 mph	59,050 ft / 1,242 miles
E-2A WEAPONS		Two NR-30 30MM cannons, one on each side of the lower forward fuselage.			

This experimental aircraft was created for two reasons; firstly, to be the first to achieve unprecedented speeds and altitudes that were envisaged for a future fighter, and secondly, to develop an interceptor capable of swiftly reaching the stratosphere and taking down an enemy bomber, or reconnaissance aircraft, such as the U-2. As a result of the new combined propulsion system, the E-2 was designated E-50. The objective was to achieve maximum speed and altitude while restricting the jet's firepower to two NR-30 cannons.

On January 9, 1956, the aircraft was ready for its maiden flight. The experienced Belarusian-born test pilot Valentin Grigorievich Mukhin took the E-50/1 to the skies for the first time, and all was well. However, on July 14, 1956, on one of the subsequent flights, the E-50's engine flamed out shortly before landing. Nevertheless, Mukhin managed to land the aircraft, at high speed, on runway 4 with its landing gear extended. The jet was propelled onto the grassy area next to the runway and came to rest after about 200 meters. Fortunately, Mukhin received only minor injuries. However, the aircraft was severely damaged due to the crash landing. Before the crash, 18 flights had been completed on the E-50, including three flights with the S-155 rocket engine.

On September 6, 1956, the MAP (Ministry of Aircraft Industry) instructed MiG to repair the prototype E-50, which had crashed landed, and to construct a second prototype, which would be flight tested together with the Air Force. Furthermore, the MAP ordered the production of 30 E-50A jets powered by the Tumansky R-11-300 engine at Gorky plant No. 21. The 30 E-50s would be transferred to the Air Force in the middle of 1957.

While Mukhin was in the hospital, recovering from his injuries, the second experimental prototype, E-50/2, was constructed. It made its first flight on January 3, 1957, with Valentin Petrovich Vasin, from the Flight Research Institute (LII) in the cockpit.

From the start, the E-50 proved to be quite unstable. In the subsonic speed range, vibrations were so strong that the plane tilted to its side. Nonetheless, Vasin was an outstanding pilot with a ton of experience, and the vibrations did not faze him. Once the plane flew through the sonic barrier, the vibrations ended, and the jet became stable. Vasin then pulled back on his stick as the E-50 rose to a height of 29,500 ft. (9,000 m). On reaching that height, he turned on the rocket engine, located in the tail of the aircraft, and just two minutes later, he remarkably reached an altitude of 65,600 ft. (20,000 m). In those two minutes, the rocket engine on the E-50 burned three tons of fuel and 'oxidizer' (a chemical that initiates combustion in other materials). After repairing some slight damage to the oxygen system, Vasin took the

E-50 to the astounding altitude of 84,000 ft. (25,600 m). Following the conclusion of factory tests, the aircraft was handed over to the LII for additional study on engine operations at high altitudes.

During that incredible test flight, Vasin attained a maximum speed of 1,530 mph (2,462 kph), reached a range of 280 miles (450 km), and the time it took him to get to 90,000 ft. was just 9.4 min. That flight wasn't Vasin's only achievement; on June 17, 1957, he broke the USSR 'Double Mach' speed record by reaching Mach 2.33, or 1,528 mph (2,460 kph). For his flights on the E-50, Vasin was awarded the 'Order of Hero of the Soviet Union.'

The E-50/3 was the third prototype in the E-50 series. This version was the only one of the three armed with two 30 mm NR-30 cannons and an ASP-5N optical gunsight. Its cannons were situated on the sides of the fuselage, just under the leading edge of the wings.

In August 1956, production of the E-50/3 commenced. During the development of this aircraft, efforts were made to rectify flaws discovered during the functioning of the prototype. However, it was decided to end production of the plane on April 17, 1957. Subsequently, it was sent for factory testing, where it completed nine flights.

The E-50/3 was significantly different from the '1' and '2' models. It had a somewhat longer nose portion that housed more avionics. The air intake was sharper looking, and the new cone was sleeker. Overall, the new jet appeared more streamlined than the previous two versions. Regarding the E-50/3's fuel system, it was able to supply the flight propulsion unit more speedily and efficiently, due to an increase in the supply of 'hydrogen peroxide' to the turbine pump. The upgraded system enabled the E-50/3 to supply and extra 229 lb. (104 kg) of fuel.

This is the third type, the sleek E-50/3, which was painted grey. (courtesy airwar.ru)

This is a side shot of the E-50/3. (courtesy airwar.ru)

This is another overview of the E-50/3, this time with its canopy fully opened. (courtesy airwar.ru)

Moreover, the new E-50/3 was equipped with a steerable nose gear. The first two versions had the swiveling nose wheel.

The E-50 jets' low range of 280 miles (450 km) and the rocket engine's enormous fuel consumption were insurmountable problems. However, the E-50 provided the designers with indispensable knowledge that was eventually applied to the development of high-speed fighter aircraft.

E-4

Because it was clear that MiG's swept wing designs would not be capable of exceeding Mach 2 under operational conditions, development on a delta wing project began.

The aircraft assigned to the mission was designated as E-4 (NATO: Fishbed), which was basically an E-2 airframe mated with a 57-degree sweepback TsAGI-59 delta wing, fitted with ailerons that provided axial compensation and had rectangular slotted flaps. In addition, three brake flaps were installed in the lower part of the fuselage, two in the wing root, and the third was rotary. The fuselage and tail assembly remained the same as the E-2, and the new wing gave the jet good maneuverability.

The main landing gear wheel wells remained in the same place. However, the landing gear was modified to retract, not into the fuselage, but into the delta wing, resulting in a change in the landing gear gauge.

Regarding the fuel system, it consisted of three groups of internal fuselage tanks with a total capacity of 415 gallons (1,570 l); the aircraft could also carry one external fuel tank with a capacity of 105 gallons (400 l).

The E-4 was mainly developed to expedite flying tests of the E-5, which had the same airframe.

The seasoned test pilot with an engineering degree and a well-deserved reputation, Grigory A. Sedov, made the historic first flight of the E-4 on June 16, 1956.

A total of 109 test flights were conducted on the E-4. The jet performed tests to ascertain the minimum flight speed and spin characteristics of the E-5 airframe, as well as the effects of aerodynamic forces acting on the canopy and the effects on the aircraft with flaps deflected by 5 degrees. The results of the tests assisted the designers in finding ways of changing the flow around the wing by installing three wing fences on the wings' upper surface and two more wing fences on the lower surface of the wings, which gave the jet an increase in span. Aerodynamicists also suggested cropping 300 mm from the tips of the delta wings. The substantially modified E-4 aircraft was then transported to the LII, where it flew a series of high-angle-of-attack flights.

The E-4 was the final stage in the lengthy development of a light frontline fighter, which eventually became the MiG-21. The E-4 was equipped with a Tumansky AM-9C engine with a thrust of 7,165 lb. (3,250 kg). That was the only difference between it and the E-5 prototype.

An E-4 with ventral wing fences and twin ventral fins. (courtesy airwar.ru)

A view from the top, clearly showing the E-4's 57 degree sweptback wings. (courtesy airwar.ru)

The E-4 was an elegant looking experimental jet, as were most of the E series aircraft. (courtesy airwar.ru)

This is an E-4 with three wing fences on top of each wing. As can be seen here, the two larger wing fences curve down and under the wings. The fences assisted by changing the flow around the wings. Note the clipped wings. (courtesy airwar.ru)

TYPE	POWERPLANT	DIMENSIONS WINGSPAN / LENGTH / WING AREA	WEIGHTS Empty / Max. t.o.w.	SPEED Max Speed	SERVICE CEILING/ RANGE
E-4	Tumansky RD-9E afterburning turbojet with 7.275 lb. thrust	25 ft 4 in / 44 ft. 1 in. / 144.8 sq ft	18.739 lb. / 11.243 lb.	805 mph	53.500 ft / 700 miles
E-4 WEAPONS		Three NR-30 30 mm cannons on the lower forward fuselage. Additionally, a 16-tube rocket pod armed with 57-mm (2.24-in) ARS-57 FFARs or a 1,102-lb (500-kg) FAB-500 HE bomb could be carried on the centerline pylon.			

E-5

The E-4 was superseded by the E-5, which had the more powerful R-11-300 (AM-11) engine delivering 11,245 lb. (5,100 kg) thrust. The engine was housed inside a redesigned detachable rear fuselage.

The E-5 also featured a larger jet exhaust and an additional PN-34-type motor-driven hydraulic pump. In addition, it was fitted with a third brake flap mounted beneath the fuselage. It also had jet pumps that cooled the engine section, two different hydraulic systems (one reserved for hydraulic power steering of the control surfaces), and it had less pointed wingtips.

On January 9, 1956, the E-5 prototype took to the skies for the first time with test pilot V.A. Nefedov at the controls. However, just one month later, on February 20, after seven successful flights, the engine caught fire during ground tests and was destroyed. The E-5's airframe was also damaged and was sent back to MiG for repairs. It returned to the test program again on March 26.

The flight tests indicated that the E-5 had outstanding flight characteristics. It also had a substantially faster vertical speed than the E-2. For the first time, a delta wing fighter reached a speed of 1,124 mph (1,970 kph). The E-5 experimental had a ceiling of 57,900 ft (17,650 m). Moreover, the E-5 reached a height of 16,400 ft (5,000 m) in just 1.6 minutes. The aircraft's takeoff weight was 9,795 lb. (4,443 kg)

The AM-11 engine in the prototype E-5 was blisteringly fast. In fact, it was about 435 mph (700 kph) faster than the AM-9 turbojet in the E-4 prototype. However, on the flip side, the AM-11 continued to suffer several performance issues, which had to be ironed out by the manufacturer. From March 26 to May 19, eight more flights were made on the E-5, the last of which saw the R-11-300 turbojet breaking down again. The engine was replaced, but things were not looking good for the Tumansky R-11-300. More engine failures occurred on May 29 and August 7. Finally, on September 22, MiG received a replacement engine, but on October 9, the day of the aircraft's 30th flight, the engine's fuel injector fastening bolts snapped.

That was it. Artem Mikoyan had had enough. From October 18, he banned all flights on aircraft equipped with R-11-300 engines, including the R-11-300 powered E-5. However, there was a silver lining. While the engines were being overhauled at Tumansky's facilities, modifications were being performed to the E-5.

Between October 1956 and February 1957, work was carried out to lengthen the E-5's forward fuselage by 400 mm to install the R-11-300 engine with an enlarged prechamber (jet igniter) on the aircraft, which caused the center of gravity (COG) to shift rearward by 5%.

The E-5 was back in the air again on March 23, 1957, and the jet performed 65 flights before another mishap occurred. After the 65th flight, the test pilot reported that the aircraft experienced stabilizer shaking. As a result, all flights were banned once again until this latest problem could be resolved.

In the workshop, the E-5 was modified to strengthen its stiffness and decrease its propensity for spontaneous oscillations. Additionally, many E-5 components were replaced since their technical service life had ended during the time it was being serviced. The E-5's tests ran from August 22 to December 1, 1957.

It was back to the skies for the E-5 on January 10, 1958. This time it seemed that it had left all its troubles behind. There were no more problems with the engine or the stabilizer. The E-5's rigidity increased significantly, according to test pilots. The plane felt more solid and sturdy.

During factory testing, the E-5 performed 98 flights. When all tests were finished, the aircraft was officially designated the MiG-21(NATO: Fishbed-A).

TYPE	POWERPLANT	DIMENSIONS WINGSPAN / LENGTH / WING AREA	WEIGHTS Empty / Max. t.o.w.	SPEED Max Speed	SERVICE CEILING/ RANGE
E-5	Tumansky R-11-300 afterburning turbojet with 7275 lb. thrust	25.42 ft / 44 ft. 1 in. / 247.57 sq ft	7,729lb. / 11,464 lb.	1,224 mph	57,415 ft / 827 Miles
E-5 WEAPONS		Three 30 mm NR-30 cannons, 60 rounds per barrel.			

An E-5 taxiing before taking off for another flight test. (Courtesy airwar.ru)

Three different views of the E-5. (courtesy airwar.ru)

An E-5 front and center. (courtesy airwar.ru)

E-50A

At a summit in Geneva, Switzerland, in 1955, US President Dwight Eisenhower suggested an 'open skies' proposal in which the US and the USSR would be allowed to undertake reciprocal aerial inspections of their nuclear installations and launchpads. Nikita Khrushchev, the Soviet leader, flatly refused the request, upholding the Soviet stance of opposing international inspections in any form.

At the same time, Khrushchev declared that the Soviet Union had built a considerable number of intercontinental ballistic missiles, causing the US administration to explore alternative measures of gauging the Soviet Union's nuclear program progress.

Thus, these fears by the US spawned the U-2 spy plane project. The U-2 was a very formidable plane. Its service ceiling was 70,000 ft., an unimaginable figure at the time. The US authorities believed that the U-2 could conduct overflights of the USSR without being detected by ground radar. That was extremely important, as any unauthorized invasion of another country's airspace was considered an act of war. All flights over the Soviet Union were operated by the CIA (Central Intelligence Agency).

In the same year that Khrushchev rejected international inspections of the USSR, the U-2 took its first flight on August 1, 1955. On July 4, 1956, the CIA conducted the first U-2 flight over Moscow and Leningrad. From then on, the flights would continue irregularly for the following four years. It was later proven that the Soviets did detect the planes on radar. Then, on May 1, 1960, the United States lost a U-2 spy plane over the Soviet Union, with CIA pilot Francis Gary Powers at the controls. Thus the United States admitted to operating spy flights over the Soviet Union.

On September 6, 1956, Gorky Plant No. 21 was entrusted by MAP with developing and building a fighter interceptor with the designation E-50A in limited series production.

The plan was to build 30 interceptors against high-altitude enemy reconnaissance and bombing aircraft, and transfer them to the Air Force by mid-1957. The reason for the urgency was the new threat posed by the Lockheed U-2 spy plane, which had breached the Soviet Union's airspace for the first time the previous month.

In the fall of 1957, the Air Force received the first E-50A prototype. Despite being based on the E-2A, the E-50A was powered by a more powerful engine, the Tumansky R-11E-100 turbojet, with 11,245 lb. (5,100 kg) thrust. The E-50A (or MiG-23U, where 'U' is the initial letter of the term 'accelerator/booster' in Russian). The new jet was designated as the E-5 'Type 64.'

The E-50A (MiG-23U) was armed with two 30-mm HP-30 cannons with barrel compensators and 60 rounds each. It was also planned to equip the interceptor with two ORO-57K units for firing 57-mm NAR (unguided rockets), but it never got there. Regarding weights, the takeoff weight of the E-50A reached 20,500 lb. (9,300 kg), and the fuel weight another 8,720 lb. (3,955 kg), a total of just over 29,000 lb. (13,100 kg), or almost triple the E-5's takeoff weight.

The E-50A contained a powerful S-155 type rocket accelerator with 8,380 lb. (3,800 kg) output and a tank holding 242.5 lb. (110 kg) of turbine pump fuel in the form of hydrogen peroxide, which was placed at the root of the tail. If that was not enough, the E-50A was also equipped with a long movable sturdy spindle-shaped container with a tank that contained 4,540 lb. (2,060 kg) of concentrated 'AK-20' nitric acid (an oxidizing agent). Within the tank, there was also TG-02 hypergolic aviation kerosene, which was rocket fuel that would ignite spontaneously on mixing with another substance. All that fuel was situated underneath the aircraft's fuselage, which made the E-50A look like an impregnated insect and a dangerous one at that.

The E-50A could have been a capable aircraft; however, it had the same complications as its predecessor, the E-2A. Due to issues with the R-11E-100 engine, plans for manufacturing the E-

50A had to be altered. Delays, and other issues, caused production to be significantly interrupted.

Consequently, in 1957, two issues rang the death knell for the E-50A project. They were the abundance of interruptions caused by the R-11E-100 engines and Tumansky's prediction.

After prolonged testing with several other MiG prototypes, the Government Ministry decided that rocket engines were not the way to go. Tumansky, the maker of the R-11E-100 turbojet engines, convinced the Government of that. He said that more powerful lightweight jet engines in the future would be faster and more efficient. Therefore, rockets were not the way to go. That was good enough for the Government. The result was the premature ending of the E-50A project.

In the end, instead of 30 E-50A jet fighters for the Air Force, only one interceptor was handed over to the military, two more were in storage, and eight E-50As were awaiting assembly.

As for the Lockheed U-2, it continued its forays over Soviet airspace for another three years until the shooting down of Francis Gary Powers' U-2 in 1960.

The E-50A prototype, with its unique belly pack. (courtesy airwar.ru)

THE E-6 PRE-PRODUCTION JET

Mikoyan was quite interested in rocket propulsion and the Sputnik program, which was very popular in the Soviet Union. However, Mikoyan was also a realist. He noticed that the first Sputnik satellites, whose undeniably spectacular successes gave many military and Government officials, as well as the pilots and engineers at MiG, the wrong idea of the absolute superiority of rockets over the perceived obsoleteness of turbojet engines. Instead, they thought that rocket propulsion was the way forward for the Soviet jets.

Mikoyan would show unwavering persistence and iron hardness to achieve his company's goal of creating the perfect jet fighter, whether with turbojets, rockets, or both. In one meeting at MiG's offices with supporters of rocket power, he said that they shouldn't forget that combat aircraft would continue to play their role in the world for a very long time and that the MiG company would not base its reliance on rocket power alone. After that meeting, the advocates of rocket power agreed with Mikoyan. The company would work together to make the best fighters on earth.

The next fighter on the agenda would bring fighter aircraft development in aviation and missile technology even closer; it was the E-6 delta wing project.

The construction for the future MiG-21 began after the results of the first experimental aircraft's performances were assessed. Test pilot Nefyodov conducted the flight tests with this prototype fighter, designated the E-6/1. 'E6' was the model number, and '1' signified it was the prototype. A Tumansky R11-300 turbojet engine powered the E6/1, with a max. thrust of 8,378 lb. (3,800 kg) and 12,655 lb. (5,740 kg) with afterburner.

The R11-300 engine, whose design and development was carried out from May 1953 to January 1955 under the famous Soviet aircraft engine designer Alexander A. Mikulin, was put into widespread production. The engine was mainly intended for MiG's new delta wing, E-6/1, and some Yakovlev aircraft.

The E-6/1 was developed from the MiG-21 (E-5) series aircraft by installing the new R11-300 engine and improving the front fuselage's aerodynamics by using a two-position nose cone and a thinner inlet lip. The traditional circular inlets with blunt lips used on previous MiGs, such as the MiG-15 and MiG-17, performed admirably up to Mach 1.4. At more incredible speeds, however, the blunt lip would generate shock waves, increasing drag. The problem would be solved with a narrower lip.

The E-6/1 prototype differed from the E/5 by a slimmer fuselage, an improved cabin interior, an arrangement of the air intakes for better engine cooling, and a drogue parachute at the aircraft's rear.

The wings still had three wing fences each, and the wing tips were clipped in the same manner as the E-4/2. However, the E-6/1 was operationally dissimilar to the E-5. The E6/1 was no longer an experimental aircraft but a prototype for the new jet fighter's intended serial production, which would be the MiG-21F.

THE E6/1 TRAGEDY

Vladimir Nefyodov was a competent test pilot. That was why he was entrusted with the E6/1 flights. On one of those E-6 flights, Nefyodov reached the target speed of Mach 2.05 while flying at the height of 41,000 feet (12,500 m). Then, tragedy struck on his seventh flight, which took place on May 28, 1958. Nefyodov flew the E-6/1 11 miles up in the middle of the stratosphere. Suddenly, he felt an explosion followed by the engine flaming out and a sudden fall in hydraulic pressure at 59,000 feet (18,000 m). The order to eject was then given. However, Nefyodov was not ready to eject; he wanted to try and save himself and his plane. He reported to base that the aircraft was experiencing severe fluttering. Nefyodov didn't know what caused the flameout, but being an accomplished test pilot, he was still thinking of the project. He intended to rescue himself, the plane, and the data that the measuring devices had gathered up to the time of the explosion. The engineers and designers would find tremendous value with those items in their possession.

The ill-fated E6/1 which was flown by the experienced test pilot Anatoly I. Nefyodov. (courtesy airwar.ru)

Nefyodov tried a risky move while 'flying' back to base in the crippled aircraft. He descended in a spiral and repeatedly attempted to restart the engine, but it would not start. As a test pilot, more than any other pilot, Nefyodov had to be prepared for these types of situations. He decided that he would have to land the fighter without engine power. That was not the only option available; he could have ejected and left the disabled jet to its consequences so he could fly another day. Not Nefyodov; he would prefer to save the plane and himself rather than be a quitter. With steel eyes and a firm grip on the yoke, he was ready to face the consequences.

He saw the runway, barely a few hundred feet below him. He had calculated everything to perfection and was sure he could do this. After the initial explosion, the control system should have instantly switched to the backup electrical system. However, with the loss of pressure, the speed of the stabilizer shift was substantially slower than the hydraulic shift. That spelled danger. He controlled the jet with what little hydraulic pressure he had. The runway was before him, and he was almost there. Suddenly, just two meters above the runway, the E-6/1 pitched down and collided with the runway, shattering the nose gear. It flipped over and tumbled for another 600 feet before coming to a halt on a cement surface with its rudder shearing off before bursting into flames.

The firefighters, ambulance personnel, the MiG engineers, aircraft mechanics, and whoever else was there; all rushed to the plane to free the pilot, who was trying to escape from the burning trap all by himself. With a broken spine, a broken rib, a punctured lung, and a badly burned back from the burning seat padding that stuck to his back, rendering him unimaginable pain, he had only one question for the doctors; would he still be able to fly? Unfortunately, Nefyodov succumbed to his injuries that very night.

After the crash of the E-6/1, a special commission was set up to find the cause of the tragic accident. It found that not one but three separate issues caused the tragedy.

The first was caused by the inlet device's unstable operation. When the intake air was compressed in the channel to a level that

Another photo of the E-6/1 which crashed and took the life of Anatoly I. Nefyodov. (courtesy airwar.ru)

greatly exceeded its penetrability, the extreme temperatures generated at twice the speed of sound led the gasoline in the launch tank to vaporize while in flight. As a result, the entire system was saturated by gasoline vapors, preventing the engine from starting. The commission found that the direct cause of the accident was the failure of the primary hydraulic control system and the slow transition to the backup electrical branch at the most critical stage when the aircraft was just two meters above the runway.

Subsequently, all the issues were successfully resolved. The internal aerodynamics of the inlet were experimentally verified in special wind tunnels. The three-position cone control of the inlet gear operated automatically after the redesign, depending on the flight mode, and could be operated manually in an emergency. In addition, a side-mounted anti-pump door and an additional intake door were added to the inlet duct.

With the loss of the E-6/1, the flight test program continued using the two remaining models; E-6/2, callsign: 'Red 22' and E-6/3, callsign: 'Red 23'. The flight trials were made between 1957 and 1958.

E-6/2

Following the loss of the E-6/1 and its pilot, the flight test program was restarted with the new E-6/2 variant. The E-6/2 was a MiG-21, s/n 65310108, converted into an E-6/2 fighter.

The E-6/2, like the E-6/1, was powered by a Tumansky R-11F-300 engine providing a maximum thrust of 8,600 lb. (3,900 kg) and 12,680 lb. (5,752 kg) with afterburner. The new engine proved an excellent match for the E-6/2. During flight tests, the E-6/2 reached a height of 41,000 feet (12,500 m) and attained a speed of 1,572.90 mph (2,531.3 kph), or Mach 2.05.

The E-6/2 entered the factory tests on August 26, 1958. Then, it took to the skies for its first flight from LII airfield on September 15, 1958. By November 5, it had already completed 15 flights.

There were some differences between the E-6/1 and E-6/2. The E-6/2 had its flaps expanded and its wingtips cropped. That was done to allow the installation of APU-13 type pylons for the K-13 air-to-air infrared 'Atoll' missiles as well as replaceable firing rails. The test program aimed to find the most suitable position for launching these A-A missiles. In other changes, the E-6/1's three-wing airflow fences were replaced with a single smaller airflow fence on the outboard wing of the E-6/2. The E-6/2 variant also had a revised tail section with the tail lowered to the fuselage axis. Also, an air intake with a thin edge was fitted at the front, which improved the plane's aerodynamics.

Finally, the wing experienced a remarkable transformation. At the base of the wing, the leading edge sweepback was 57 degrees. However, from about the midpoint of the wing to the wingtip, it was lowered to 48 degrees. That was not done for aesthetic purposes but out of necessity. Placing the pylons, launch bars, and missiles on the wings presented several difficulties. The engineers determined that the only viable solution was to add extra room to the ends of the wings to accommodate the launchers and missiles. Engineers expanded the wing area, resulting in the 48-degree sweepback.

After the E-6/2 completed 46 flights, the program came to an end. Fifteen of the flights were undertaken at LII. The other 31 flights were conducted at Krasnovodsk Peninsula in western Turkmenistan, a favorable location for supersonic jet flight because it is a vast, sparsely populated area.

E-6/3

On November 20, 1958, the E-6/3, 3rd prototype, was shipped to the manufacturer for testing. At the plant, the E-6/3 had anti-surge controlled flaps installed, which were added to bypass air

The E-6/2 on the mat. (courtesy airwar.ru)

into the atmosphere to preserve air intake stability when the engine was throttled over Mach 1.5 during flight.

The E-6/3 began flight testing in December 1958 and made its first flight on Christmas Day 1958, piloted by the famed test pilot Georgy K. Mosolov. Like the E-6/2 prototype, this aircraft was based at the Krasnovodsk Peninsula.

In 1959, the E-6/2 and E-6/3 passed joint State tests performed by test pilots Aleksandr Fedotov and Konstantin K. Kokkinaki at the State Research Institute of the Air Force. The E-6/2 and E-6/3 completed sixty-one factory test flights with no incidents. Kokkinaki was confident about the two jets and said that any pilot could fly them easily. At the same time the tests were going on in 1958, serial production of the MiG-21F (NATO: Fishbed-B) was launched. The MiG-21 was finally ready.

E-66 / E-66T / E-66A

The USSR began large-scale serial production of the MiG-21F (previously E-6/3) and sent it to Soviet AF units in the late 1950s. The plane was then progressively integrated into the arsenal of the Warsaw Pact's republics in the first half of the 1960s and, from 1963, into the Air Forces of India, Finland, and other states.

After completing the factory tests, the Soviet Union had the MiG-21F enter several FAI (Fédération Aéronautique Internationale) record disciplines that the United States previously dominated. The first entry was a lightened E-6/3 (Mig-21F) which flew under a bogus designation ('E-66'). The jet was primarily used to break speed and altitude records. It was temporarily equipped with a more powerful Tumansky R-11F-300 engine, but registered as using an 'R-37F' engine, which was the factory name of the R-11F-300. On October 31, 1959, Mosolov established the absolute world speed record over a 9/15-mile (15/25-km) course at a speed of 1,483 mph (2,388 kph). One run was timed at 1,556 mph, (2,504 kph) about Mach 2.356. Then, on September 16, 1960, Konstantin K. Kokkinaki flew the MiG-21F (E-6/3) at a speed of 1,334 mph (2,146 kph) on a 61.2-mile (100 km) circuit. Both records were reported to the FAI as having been established by an aircraft with the 'E-66' designation.

At the beginning of 1961, the MiG Design Bureau released the E-6T/1 (the E-6T-1 became the prototype of the MiG-21F-13). This time MiG made significant modifications to the jet to enable it to achieve the highest flight altitude. It was equipped with an R-11F2-300 engine and a U-21 booster with a 45-gallon (170 l) SZ-20M5A Liquid-Propellant Rocket Engine (LRE) positioned below the fuselage delivering 6,613 lb. (3.000 kg) thrust. After the redesign, the aircraft was designated 'E-66A'. The E-66A achieved a major world record.

On April 28, 1961, Mosolov flew the E-66A and attained an altitude of 113,891 ft. (34,714m). However, the flight was not without incident. The main engine shut down at the height of 73,820 feet (22,500 m), and the climb was completed using the rocket engine.

Images by Newresid stephanelhernault@yahoo.fr

The E-6/3 was the prototype MiG-21F, where 'F' stood for 'Uprated', NOT fighter! (courtesy airwar.ru)

The E-6/3 on the apron. The progenitor of the Mig-21F. (courtesy airwar.ru)

Cockpit of the MiG-21F (E-6/3) (courtesy airwar.ru)

Later, as the airplane approached the highest point of its trajectory, it began to spin gently around the longitudinal axis. The 'E-66A' was on the brink of stalling and going into an inverted spiral, but Mosolov corrected the situation. The altitude record stood for more than a decade. In 1973, Mosolov lost it to the E-266 (the MiG-25 'Foxbat'). Mosolov needn't worry; he set six European aviation records, of which three are absolute and three are USSR/Russian absolute records.

1st GENERATION SERIES

THE START OF MASS PRODUCTION

MiG-21

Taking a few steps back to 1956, on June 11 of that year, the MiG facility in Tbilisi, plant no. 31, Soviet Georgia, was commissioned to build a small series of Mig-21 jets. The MiG-21 (Fishbed A) was the production version of the 'E-5'.

The Mig-21 was regarded as having the finest supersonic interception capabilities; however, the fact that its low subsonic payload capacity severely hampered its ground assault capabilities was overlooked.

Its armament consisted of two NR-30 guns under its fuselage and two air-to-air K-13A (NATO: Atoll) missiles with an infra-red guidance system carried under its wings.

Undoubtedly, this early version of the MiG-21 lacked power, and the critical weight factor necessitated the removal of the left cannon. However, in 1959, the turbojet engine's thrust was somewhat boosted. MiG installed the new R-11 F-300 on the Mig-21, and the aircraft's thrust increased to 12,680 lb. (5,750 kg).

In the end, only seven MiG-21 fighters were built at Plant No. 31 in Tbilisi before manufacturing was halted. Construction was switched to the more advanced 'E-6', MiG-21F. Subsequently, the seven Tbilisi-built MiG-21s served as test platforms.

MiG-21F

Image by bagera3005, Deviantart.com

The MiG-21F (E-6/3), factory designation 'E-6T', 'Type 72' (NATO: Fishbed B), was a single-seat day fighter. This model was found to have solved most of the problems of the original MiG-21 and went into large-scale production at plants no. 21 in Gorky, Moscow, and no. 31 in Tbilisi, Georgia. The Russian and Georgian plants both manufactured 30 MiG-21Fs in 1959 and another 63 in 1960. In total, 99 Mig-21Fs were produced.

The first production aircraft was given the designation MiG-21F (the letter 'F' stands for 'Forsirovanny,' which translates as 'forced,' referring to the increase in thrust). They began to equip the Russian Air Force (Voenno-vozdushnye sily Rossii, 'VVS') in the fourth quarter of 1959. Shortly after the launch of this series, the vertical tail chord was increased. This modification became standard on all MiG-21Fs built after that.

The MiG-21F, whose capacity was limited to interception operations on clear days only, was equipped with a simple telemetry radar located in the air intake cone. The VVS retained only the front right-side NK-20 cannon as standard operational armament.

A MiG-21F fully loaded with armament. (courtesy airwar.ru)

A good view from above of the MiG-21F (E-6/3) (courtesy airwar.ru)

Subsequently, the MiG-21F was designed to carry 571 gallons (2,160 l) of fuel in six internal fuel tanks and was powered by a 12,655 lb. (5,740 kg) thrust Tumansky R11F-300 turbojet engine.

Regarding armament, the early production aircraft were equipped with one NR-30 and two NR-23 cannons. Yet, subsequent aircraft only carried two NR-30 cannons, with 60 rounds for each gun installed on opposite sides of the Mig-21F's lower fuselage. In addition, on the bottom surface of each wing console was a BD3-58-21 pylon from which two 1,000 lb. (500 kg) bombs or UB-16-57U blocks with S5 unguided missiles could be carried. Furthermore, a 130-gallon (490 l) drop tank was placed under the aircraft's fuselage (see photo below).

When the MiG-21F was being developed, ground attack played a secondary role, and flying performance, particularly air combat performance at all altitudes, was the primary concern. Despite slight increases in engine thrust, the necessary performance was not at the level that MiG wanted. Part of the issue was unquestionably an overweight airframe; nevertheless, despite the MiG OKB's skill in estimating weight, the MiG's fundamental heaviness remained an issue. There were also other issues with the MiG-21F.

For instance, in 1962, when the Indian government purchased MiG-21Fs for assessment purposes, the IAF opted to reinstall the left cannon on most aircraft despite the weight penalty. The MiG-21F provided superb agility and handling for Indian pilots. The MiG pilots enjoyed excellent sight, but despite its inherent simplicity and ease of maintenance, the MiG-21F had several flaws, including limited autonomy, inadequate weaponry, and insufficient braking capacity. According to the Indians, the K-13 infrared air-to-air 'Atoll' missile was only effective against docile targets in clear weather, and only when fired fully from behind the enemy aircraft.

The MiG-21F was also produced in Czechoslovakia under license. The main difference between the Soviet and Czech versions of the MiG-21F was the removal of the rear cockpit windows on the Czech jet.

Another view of theMiG-21F. Note the 130 gallon drop tank hanging underneath the airframe. (courtesy airwar.ru)

TYPE	POWERPLANT	DIMENSIONS WINGSPAN / LENGTH / WING AREA	WEIGHTS Empty / Max. t.o.w.	SPEED Max Speed	SERVICE CEILING/ RANGE
MiG-21F Type 72	Tumansky R-11-300 (also known as R-37F) turbojet with 12.125 lb. thrust with reheat	25.42 ft / 44 ft. 1 in. / 247.57 sq ft	10.625 lb. / 15.100 lb.	1.320 mph	62.335 ft/ 945 miles
MiG-21F WEAPONS		Three 30 mm NR-30 cannons, with 60 rounds per barrel. Two 1,000 lb. (500 kg) bombs on underwing pods, or UB-16-57U blocks with S5 missiles, or two K-13/R13 AA Atoll missiles.			

DESCRIPTION OF THE MIG-21F

The MiG-21's fuselage had a conical center body, a 'semi-monocoque' airframe (structural pieces running horizontally down the plane to help hold the frame together), a narrow nose air inlet, and a conical center body. The tail unit control rods were enclosed by a dorsal fairing that stretched from the back of the upwardly hinged cockpit canopy to the bottom of the tail fin. The canopy was jettisoned with the ejector seat for increased pilot protection during an emergency. Cooling air intakes were positioned at the back of the fuselage, and there was a leaf-shaped reheat pipe with two flame holders. All the MiG's 620 gallons (2,340 l) of fuel were stored within the fuselage along with a tiny water-methanol tank. Moreover, a pod below the fuselage could accommodate a vast finned fuel tank with a capacity of 170–200 gallons (640–750 l) of fuel, a missile, or a bomb pack. There was a small ventral air brake behind the pod, and for added security, stowed in the lower port side of the rear fuselage was a drogue chute.

E-6T/1 /2 /3

After a brief lull in production, the 'Znamya Truda' factory delivered the 99th MiG-21F. Production once again resumed with the newest variant, which was the E-6 prototype armed with the K-13 Air-Air 'Atoll' missiles and designated E-6T/1. The serial aircraft derived from it bore the designation MiG-21F-13.

In the fall of 1959, many MiG-21F aircraft were dispatched to test the brand-new K-13 'Atoll' air-to-air missile system. 'Red 31' was one of the MiG 21s taking part in the testing. That specific jet was at the forefront of the K-13 missile testing and was outfitted with APU-13D launchers. Thus, 'Red 31' was designated E-6T/1. The E-6T/1 was also involved in other tests. Actually, 'Red 31' was also the jet that made history on October 31, 1959, when test pilot G.K. Mosolov broke the speed record (then designated E-66).

The E-6T/2 was dispatched to the unpaved Tretyakovo airport in 1961, situated in Lukhovitsy, Moscow Oblast. Its task was an easy one. The E-6T/2 was equipped with non-retractable round and rectangular skis, and tests were done on dirt strips to determine if mounting the skis on Soviet Air Force aircraft was viable. It was not.

Later, the next test for the E-6T/3 MiG-21F-13 ('Fishbed C') was to be equipped with canards. According to theoretical assumptions, installing canards to a fast jet would increase longitudinal stability at transonic speeds. It was E-6T/3's job to confirm that theory. While traveling at subsonic speeds, the canards were free floating. But, to avoid mishaps, the canards were locked during supersonic speeds. MiG's OKB conducted canard tests on the 'E-6T/3' (Mig-21F-13). However, OKB never considered installing canards on the early MiG fighters. Canards were intended to be installed on future aircraft such as the E-8. However, the project did not advance beyond the prototype stage due to issues with the jet propulsion systems. In the end, neither the 'E-8' nor the 'E-152M' heavy interceptor took to the skies. The 'E-6T/3' prototype was the sole aircraft equipped with canards. It completed a total of 58 test flights before MiG discontinued the canard project in July 1962.

MiG-21F-13

The first generation of MiG-21 jets were fighters with modest equipment used primarily for daytime operations and in good weather conditions.

With the success of the K-13 infrared air-to-air missile tests, on February 2, 1960, the Soviet Union's Council of Ministers issued an order to begin mass production of the K-13 infrared homing missile. The selected factories to produce the K-13 were 43 in Moscow and 485 in Kyiv. The plane chosen to be equipped with the K-13 missiles was the MiG-21F. By outfitting the 'F' version with the K-13, it was redesignated 'MiG-21F-13' and given the title 'Type 74'. The 'F-13' would be a much more formidable fighter when armed with two K-13 missiles on its underwing pods than the MiG-21F variant.

The MiG-21F-13 was also the first second-generation MiG-21. In addition, it was the first to be built in substantial numbers. It was also the first to experience many modifications. Beginning with MiG-21F-13 serial number 0815, the NR-30 cannon on the left of the 'F-13's' fuselage was rendered unusable due to its heavy weight. However, in case the cannon was needed, the space necessary for it was retained, and a simple fairing covered the muzzle of the suppressed weapon.

The early MiG-21F-13 had no outward differences from the MiG-21F. However, beginning with the 115th production model, the vertical stabilizer on the 'F-13' model was widened and shortened, and the canopy framing was slightly modified. Furthermore, the fuel supply in the internal wing tanks was increased by 53 gallons (200 l). After these modifications, NATO didn't miss a beat and dubbed the newest MiG variant, 'Fishbed C.' The late-series versions of the MiG-21F-13 were produced in the USSR and under license in Czechoslovakia between 1962 and 1966.

The MiG-21F's onboard equipment lacked an autopilot; however, on the Mig-21F-13, a KAP-2K autopilot was fitted. Also installed were an ARK-10 radio compass, an RV-UM radio altimeter, an R-802V multichannel radio, an MRP-56P radio beacon receiver, a SOD-57M transponder, SRO-2M IFF (Identification Friend or Foe), and a Sirena-2 Radar Warning Receiver with a maximum range of 138 miles (222.2 km). Furthermore, the SRD-5 radar rangefinder was superseded by the SRD-5M radar rangefinder on late-production MiG-21F-13s.

TUMANSKY R-11-300 TURBOJET

The R11-300 (formerly designated 'AM-11') turbojet was conceived by the renowned engine designer A. A. Mikulin. Initial samples were subjected to tests (from May 1953 to January 1955). Numerous upgrades to the R11-300 turbojet engine were planned for MiG and Yak planes. However, it was the MiG-21 that wrote the story of the R-11. It was the jet engine that immortalized the MiG-21 and vice versa. The MiG-21 and the R-11 were inseparable.

The Tumansky R-11-300 engine was a modulated two-spool axial-flow turbojet with afterburner. The compressor comprised a twin-section, twin-shaft design with a three-stage low-pressure section and a three-stage high-pressure section. There were ten circular combustion liners in the combustion chamber. The turbine also consisted of two sections and two shafts. The front turbine and second compressor segments shared a shaft, whereas the rear turbine section shared a shaft with the first compressor section.

In January 1956, the first aircraft to use the new R-11-300 (AM-11) turbojet engine was the MiG-21's progenitor, the E-2A/1.

In total, about 20,000 R-11-300 engines in various configurations were built. In addition, a Chinese copy of the R-11-300 was also produced and fitted on the J-7 fighter.

The Tumansky R-11-300 Turbojet with afterburner. (Varga Attila)

EXPORT MIG-21F-13 'FISHBED C'

Plant no. 30 produced and exported the 'Mig-21-F-13' in 1962. All exported day fighters were the MiG-21F variant.

In 1959, MiG-21 jets were first inducted into the fighter units of the Soviet Union's military. Once that was completed in 1961, pilots from the Warsaw Pact made the first flights on the MiG-21F-13. Then, at the start of 1963, the new MiG-21F-13s were delivered to the Air Forces of the Warsaw Pact, Finland, India, and China, followed by several other nations.

Finland was the first country outside the Warsaw Pact to receive MiG-21s. So when the MiGs were delivered to that country on April 24, 1963, Western observers got their first close view of the Soviet Union's most important fighter aircraft. Ten Mig-21-F-12 fighters (the Finnish F-13s were designated F-12) were delivered to the Finnish Air Force. The MiGs were flown to Räyskälä Airport in Kuopio, Finland, by a team of Russian pilots.

Kuopio was the base of the Finnish Air Force's 31st Fighter Squadron. The MiG-21F-12 served with the Finnish Air Force until January 17, 1986, when newer MiG variants replaced it.

As mentioned, the 'F-13' was also built in Czechoslovakia at the Aero plant near Prague. Also, in Poland, in November 1961, the first 15 Polish pilots received training on the MiG-21 in the Soviet Union, and the MiG-21F-13 aircraft were inducted into the Polish Air Force in 1963. They were presented to the public for the first time on July 22, 1964, over Warsaw.

The first MiG-21 F-13 manufactured in China and designated J-7 took flight in December 1964. License production of the 'F-13' started in Czechoslovakia in 1962.

India also manufactured a substantial number of MiG-21 aircraft. In January 1963, the first six MiG-21 F-13s, 'Type 74' arrived in Mumbai.

The MiG-21F-13 was initially delivered to the Indian Air Force's 28th Squadron. Eight Indian pilots from the Squadron received three months of training on the 'F-13' in Kyrgyzstan beginning in October 1962.

Incidentally, two of those first six Mig-21 fighters delivered from the USSR, before Indian license manufacturing of the type, were lost in a mid-air collision while flying in formation from 'IAF Chandigarh' in Punjab State in India on December 21, 1963. The pilots ejected safely.

The E-6T/3 MiG-21F-13,was equipped with the canards. It made 58 successful test flights before testing stopped in 1962. (courtesy airwar.ru)

The prototype MiG-21F-13 (E-6T) day fighter. (courtesy airwar.ru)

TYPE	POWERPLANT	DIMENSIONS WINGSPAN / LENGTH / WING AREA / HEIGHT	WEIGHTS Empty / Maximum takeoff weight	SPEED Max Speed / Landing Speed / Maximum rate of climb. m/s	SERVICE CEILING/ RANGE
MiG-21F-13 Type 74	Tumansky R-11-300 turbojet with 8,600 lb. thrust and 12,676 lb. with a.b.	23.47 ft / 44.1 ft / 247.57 sq ft / 15.76 ft.	10,979 lb. / 19,015 lb.	1,351 mph / 174 mph / 0.08 miles per second	62,335 ft/ 808 miles
MiG-21F-13 WEAPONS		One HP-30 cannon installed in the fuselage on the lower right of the fuselage. Underwing weapons: 32 NAR S-5M or S-5K unguided rockets, two NAR S-24 unguided missiles, two 50 kg bombs or two ZB-360 incendiary tanks			

MiG-21-F-13 c/n 741004, s/n 619. Delivered to East German AF in Oct. 1963. It flew with fighter squadron JG-3, and fighter pilot training squadron 15 (JAG-15). WFU Nov. 1982. The s/n '335 red' in the above photo is fake. This jet's actual s/n was 619. (Rob Schleiffert)

This MiG-21F-13 was delivered to Czechoslovak AF in Oct. 1964. It served with fighter units:11 SLP, and 5 SLP. WFU in Dec. 1988, and sold to Hungarian AF. It is Currently preserved at McClellan AFB, California. (Alan Wilson)

This Czechoslovak AF MiG-21F13 is preserved at Pardubice, s/n:0309, c/n:560309, Behind it is a camouflaged Su-7 Fitter. (Rob Schleiffert)

MiG-21F-13 delivered to the Polish AF in Feb. 1962, s/n 2015, c/n 74212015. It was based at Deblin AB from 1962-1974, then served with '2 PLM' fighter aviation regiment in 1974. From 1974-1977 it was with 'Central Training Center for Air Force Technical Specialists' (COSSTWL). Currently at Muzeum Sil Powietrznych, in Poland. (Alan Wilson)

MiG-21F-13 from above. Note the K-13 infrared Air-to-Air missile (NATO: 'Atoll') below the left wing. (courtesy airwar.ru)

This is a Cuban 'F-13' s/n 411, c/n 74211405. It was constructed in 1961, and served with the USSR Air Force, before Cuba received it. It is currently preserved at the DAAFAR Museum in Havana, Cuba. (Richard Vandervord)

This is one of the firs serial MiG-21F-13 aircraft. Note the four small wing fences, two on each wing. (courtesy airwar.ru)

This is a MiG-21F-13, c/n 741721 s/n MG-92, which was delivered to the Finnish AF in Oct. 1965. It was assigned to the 31st squadron before being modified to a reconnaissance variant. It was then assigned to Jyvaskyla-Luonetjarvi AB's Recon Squadron. As shown above, it is presently housed in the Central Finland Aviation Museum in Tikkakoski AB. (Richard Vandervord)

This is a Yugoslav MiG-21F-13 (L-12), s/n 22532, c/n 741702. It was delivered to the county's Air Force in October 1964 and served with the following units: 204th LAP, 117th LAP, and 83rd LAP. It is presently preserved at the Belgrade Aviation Museum. (Richard Vandervord)

Indonesian MiG-21F-13 c/n 74212110 s/n 2160. This F-13 was delivered to the Indonesian Air Force in 1962 and was assigned to SkU 14. It is presently kept in Yogyakarta at the IAF Museum. (Dirgantara Mandala Museum)

MiG-21F-13 Czechoslovakian c/n 161101 s/n 1101. This fighter was converted to MiG-21FR standard and delivered to the Czechoslovakian Air Force in November 1971. It was assigned to SLP unit 1. It was then stored at Vodochody AB in March 1991. In 1996 it was transferred to the March Field Museum, in Riverside, California. (Alan Wilson)

2nd GENERATION SERIES

E-6V/1 & /2

In 1961, after the MiG-21 F-13 aircraft was put into series production, OKB MiG was commissioned to develop a modified MiG-21 F-13. Subsequently, OKB decided to build two experimental aircraft on the 'F-13' series. The experimental jets were built to test the usage of dischargeable rocket boosters on JATO (Jet Assisted Take-Off) and an enhanced parachute braking mechanism during takeoff and landing. Furthermore, they were to test their performance under harsh flying conditions similar to those that the MiG-21F-13 'type 74' production series fighters may encounter. Therefore, OKB MiG gave these two new experimental jets the designation: E-6V/1 and E-6V/2.

As part of the conversion, the two aircraft had the more powerful Tumansky R-11F2S-300 engine (product 37F) installed, which delivered 8553 lb. (3,880 kg) thrust dry and 12,654 lb.(5740 kg) with afterburner.

The primary difference between the E-6V experimental versions and the standard MiG-21F-13 'type 74' design included a box with a drogue parachute located at the base of the trailing edge tail surface just under the rudder, with a conical tip that was situated beyond the trailing edge of the tail.

Attachments for two assisted-takeoff solid rocket motors were installed on the lower sides of the rear fuselage on both jets during the JATO tests. The most important one was the SU-1500, which had a 10-second thrust rating of 7,716 lb. (3,500 kg) and was mounted horizontally, although the nozzle was angled 45 degrees to thrust in a line across the aircraft's center of gravity, as was the case for all other rockets.

On July 9, 1961, during the aviation display in Tushino, test pilot A. V. Fedotov flew the E-6V/2 without the suspended missile armament, essentially taking off from the airfield area using only the rocket engines.

The E-6V/1 was further modified in the following testing phase to check the viability of takeoff from a large angle, between 16 and 18 degrees. Whereas, in regular operational service, the planes would take off and land at an angle of attack between 10 and 11 degrees. Unlike the production MiG-21Fs and MiG-21F-13s, the E-6V/1 had fixed flaps in position for only takeoff and landing. Interestingly, the external difference between the E-6V/1 and the MiG-21F-13 was only in the housing of the drogue parachute.

Flight testing of the E-6V/1 prototype in this modification was conducted in July and August 1963. The testing found that the experimental model lacked longitudinal stability at high angles of attack during takeoff and landing. Moreover, it was challenging for the pilot to see, which caused considerable difficulties during the landing maneuver, especially in its final phase. It was also demonstrated that the pilot would need to be specially trained to perform flight maneuvers at high angles of attack and that a less experienced pilot would have tremendous trouble doing so in most circumstances. Overall, it was proven that landing with an airplane modified in this manner was significantly more complex than landing with jet flaps (powered aerodynamic high-lift devices used on the wings of specific aircraft to improve their low-speed flight characteristics).

This is the second prototype P-6V/2. The use of the two experimental planes confirmed that taking off and landing in acute angles with high angles of attack was not a good idea. (courtesy airwar.ru)

The E-7/1 'Red 71' with K-13 missiles. (courtesy airwar.ru)

E-7V/1 /2 /3 & /4

The new interceptor variant of the MiG-21 earned three names; the OKB MiG designated it E-7, the manufacturer 'Type 76', and the Air Force gave it the military description MiG-21P. The designation 'P' stood for 'Perekhvatchik' or 'Interceptor.' The E-7's primary mission was to destroy air targets in the framework of air defense, its secondary role was to strike ground targets.

Three E-7 prototypes, designated E-7/1, E-7/2, and E-7/3, were equipped with a Tumansky R-11-F-300 engine and were built in a short period of time. Furthermore, all three jets preserved the fundamental and distinguishing design and appearance features of the experimental prototype E-6/3 and the serial production operating aircraft, the MiG-21F, and MiG-21F-13, on which they were based.

However, there was one significant difference with the E-6/3. The E-7s carried a TsD-30T (RP-21M Safir) (NATO: Spin Scan A) impulsion radar instead of an SRD-5M radar range finder. As a result, the air inlet had to be redesigned, lengthened, and expanded by 0.52 ft. (0.16 m) in diameter to make room for the new radar. In addition, the locator antenna unit was housed within a bigger center cone. This modification resulted in a more prominent dorsal spine and a longer forward fuselage, which gave the interceptor a more streamlined look.

The E-7/1 prototype, 'Red 71', was fitted with an identical vertical tail and keel as the front-line MiG-21F (Fishbed C). In addition, it was equipped with an ARK-54I radio compass, MRP-56P marker, SRO-2M, and SOD-57M systems.

Being an interceptor, the MiG-21P's only armament was two K-13 missiles. Both 30 mm NR-30 cannons from the aircraft were eliminated in a vain attempt to make the jet lighter. In fact, despite the removal of the guns, the E-7/1 was heavier than the MiG-21F.

At the end of the 1950s, the Soviet Air Force aimed to ensure that almost all combat aircraft could operate from unpaved airfields. Therefore, the E-7s were equipped with larger wheels.

Moreover, dismantling the barrel armament was an unfavorable modification to the MiG-21's original design, which was reflected in its combat capabilities and had to be corrected based on combat experience during the Middle East and Vietnam conflicts. Following the experiences of those two wars, MiG mounted a GP-9 suspension case with a twin-barreled 23 mm G gun-23L on the MG-21FL and MG-21PF. Furthermore, the MiG-21's 30mm guns were reintroduced in the third generation series of the MiG-21.

The E-7/2 'Red 72' interceptor was constructed in December 1959, and the E-7/3 'Red 73' in May 1960. Several adjustments were made to the designs of the two jets in order to improve their directional stability when flying at high speeds. As a result, there was an increase in the bow of the fin. Meanwhile, the side surface and thickness of the missile suspension pylons were reduced. Regarding equipment, a KAP-1 autopilot was added to the E-7s. Also, the K-13R missile was tested on the E-7/3 prototype in 1963.

The E-7/4, 'Red 74', had the upgraded Tumansky R-11F2-300 turbojet. Installing an overhead tank in the dorsal spine of the fuselage and rear wing tanks increased the fuel system's capacity. In addition, the nose of the aircraft was modified to facilitate the installation of the TsD-30TP (RP-21) anti-jamming radar.

In 1962, the E-7/4 was tested with a wheel-ski chassis to determine if skis could be used on unpaved airfields.

Joint state flight tests of the four E-7 aircraft were completed on June 30, 1961, and the results were positive. A total of 160 flights on the E-7/3 and another 154 on the E-7/4 were made during the tests. As a result, the E-7 was recommended for mass production and adoption by the Air Force. The serial production version of the E-7/4 all-weather interceptor model was to be the 'MiG-21PF'.

MiG-21PF

In 1962, the assembly lines were busy producing the newest MiG, the MiG-21 PF, where 'F' stood for 'Forsirovannyy' or (uprated). This version was a direct successor of the MiG-21F-13, powered by the more powerful Tumansky R-11F2-300, rated at 8,708 lb. dry (3,950 kg) or 13,492 lb. (6,120 kg) with afterburner.

The first production MiG-21PF, 'Type 76' (NATO: Fishbed-D), coming off the production line was s/n 76210101. It made its first flight on June 28, 1961, from the MiG factory airfield in Gorky, with test pilot S.I. Savchenkov at the controls. The flight went off without a hitch, and the following month, in July, it was displayed at Tushino's annual air show.

The MiG-21PF was the first genuinely operational version of the MiG-21 equipped with a radar in regular service. The 'PF' 'borrowed' various improvements from the experimental MiG-21P (E-7), including bigger wheels, a broader air intake, booster fixation points, and, of course, the TsD-30T (RP-21M Safir) (NATO: Spin Scan A) impulsion radar. However, the 'PF' differed from the 'P' in that it had a wide-chord tail fin, giving it a drastically different look from previous models.

The rear cockpit window was removed, and an overhead 'fuel hump' was installed in the top section of the fuselage and rear wing tanks, boosting the fuel system's capacity to 605 gallons (2,750 l). As a result, the MIG-21PF had 11 fuel tanks: seven in the fuselage, four in the wings, and one under the fuselage. The MiG-21PF was the first production version equipped with the 'fuel hump.'

The pitot tube on the MiG-21PF was also relocated from beneath the nose section of the fuselage to the aircraft's front upper right side. That was the arrangement for the pitot tube on all MiG-21s from then on.

It was Mig's policy to improve the MiG-21 continuously. Early in the MiG-21PF's manufacturing cycle, MiG discarded the aircraft's original braking parachute. It was removed because of its location beneath the back of the fuselage. The new brake chute was placed in a bullet fairing at the base of the tail fin. Subsequently, this change led to the introduction of a completely redesigned tail fin whose chord was considerably increased, which led to improved directional stability at very high speeds. Due to the more efficient positioning of the braking chute and the improved directional stability, Mig effectively killed two birds with one stone.

The MiG-21PF lacked a fixed gun, instead relying on K-13 (AA-2 Atoll) air-to-air infrared missiles or K-5 'Line-Of-Sight Beam Riding' (LOSBR) Kaliningrad K-5 (NATO: 'Alkali') laser homing air-to-air missiles. Apart from the missiles, the two pods beneath the wing could carry bombs and unguided rockets.

MIG-21PFs were built at the Gorky plant no. 21 between 1962 and 1964 for the Soviet Air Force and then again between 1964 and 1968 for export orders. The MiG-21PF entered service with the USSR's fighter regiments and the Air Force's fighter units in 1962.

TYPE	POWERPLANT	DIMENSIONS WINGSPAN / LENGTH / WING AREA / HEIGHT	WEIGHTS Empty / Maximum takeoff weight	SPEED Max Speed / Landing Speed	SERVICE CEILING/ RANGE
MiG-21PF Type 76	Tumansky R-11F2-300, rated at 8,708 lb. dry 13,492 lb. with ab	23.45 ft / 46.25 ft / 247.57 sq ft / 15.45 ft.	11,355 lb. / 20,015 lb.	1,351 mph / 161 mph	62,335 ft / 1,044 miles
MiG-21PF WEAPONS		2 x infrared homing R-3S air-to-air missiles			

This is the experimental MiG-21PF (E-7/4) with the ski-wheel chassis. (courtesy airwar.ru)

CCCP 1967

Egypt 1972

East Germany 1974

MiG-21 PF
"FISHBED D"

Images by Newresid stephanelhernault@yahoo.fr

This is a Polish AF MiG-21PF, c/n 760615, s/n 0615, of the 'Warszawa' unit. It had quite a long service life. It was delivered to the Polish AF in Jan. 1965, and served in the following units; 1965-1966: 11 PLM, 1971: 13 PLM, 1971-74: 1PLM, 1980: 39 PLM, and from 1980 to 1988 it was back with 1PLM. It is currently preserved in Warsaw-Rakowice. (Alan Wilson)

Mig-21PF of the Soviet AF, '69 Blue' code: 69 o/l, c/n: 94210601. (Mike LaChance)

MiG-21PF (RFM), c/n 760501, s/n 501. Delivered to the Hungarian AF in 1964, and served with 47 HVE. In 1977 it was assigned to 59 HVE. Wfu in December 1988. It is currently preserved in the IWM at RAF Duxford. (Richard Vandervord)

Hungarian MiG-21PF, c/n 761512, s/n 1512. Received by Hungarian AF in 1965. Served with units; 31 HVE, in 1977 with 59 HVE, until Dec. 1988, when it was WFU. Currently situated at Szolnok museum. (Richard Vandervord)

A Czechoslovakian MiG-21PF, c/n 760302, s/n 0302. It was delivered in June 1964 and operated with fighter Regiments 8 SLP and 9 SLP until it was Withdrawn From Use (WFU) in May 1989. It is preserved in the Letnany museum. (Alan Wilson)

Polish AF MiG-21PF, c/n 761702, s/n1702, delivered in July 1965. It was a part of Fighter Regiment 62 PLM until 1989, when it switched to 1 PLM. WFU in 1989. It is now on display at Muzeum Sil Powietrznych. (Alan Wilson)

The MiG-21PFL was designed for the humid South East Asian region. This is MiG-21PFL, c/n 772109, s/n 021-09, board/no. 4324. Seen here at the Hanoi Military History Museum, and preserved there since April 1996. The aircraft was delivered to the North Vietnamese AF in 1966, and operated with the 921st Regiment until 1985. (dronepicr)

MiG-21PFL

The MiG-21PFL variant evolved from the MiG-21PF. However, the 'PFL' model was adapted to Vietnamese specifications. The 'L' was allegedly an abbreviation for 'Locator' or ‘Radar’ to indicate the differing sensor suite in this variant from the regular MiG-21PF.

This MIG-21PFL was built in 1965 by 'Znamya Truda' (Moscow Aircraft Production Association) and was a 'Type 77VK' MiG-21.

The MiG-21PFL featured the same weaponry and equipment as the main 'PF' variant, but it lacked a range counter on the ARK-10 radio compass and had no radio altimeter; however, it was equipped with a KSI-2 compass.

MiG-21PFS (E-7SPS)

MiG's experimental fighter-interceptor 'E-7SPS' 'Type 94' SPS stood for 'Sduv Pogranichnovo Sloya' or 'Blowing Off the Boundary Layer.' As the phrase says, the E-7SPS was equipped with a boundary layer control blowing system (BLC). The boundary layer control is used on aircraft wings to prevent laminar and turbulent separation by removing flow of low momentum.

Work on the BLC was performed on an E-7SPS series aircraft, s/n 76210103, which was reequipped with new apparatus for the tests. During the first half of 1962, the E7SPS passed factory and state tests. In the BLC tests, it used flaps to blow off the boundary layer and used a new parachute-braking system.

That made it possible to shorten the average departure run of the MiG-21PFS' by 1,575 ft. (480 m) and its landing speed by 150 mph (240 kph). The aircraft was also fitted with two PRD-99 launch boosters to shorten the takeoff run and an R-11F2S-300 engine with modifications delivering improved injecting characteristics providing a thrust of 13,613 lb. (6,175 kg).

The E-7SPS version (NATO: Fishbed D) was put into mass production in 1964 at the Gorky plant under the designation MiG-21 PFS 'Type 94'.

The E7SPS which was put into production as the MiG-21PFS. (courtesy airwar.ru)

This is one of the few MiG-21 FPS variants that were built. (courtesy airwar.ru)

MiG-21PFM

The MiG-21 PFM (M stood for 'Modified') 'Type 94' Interceptor was the final significant version of the second generation of MiGs. (NATO: Fishbed F). This variant had all the progressive improvements specified on the PFS variant and several innovations like a separate windscreen and sideways-hinged canopy.

The canopy had a fixed front part and the central part that swung to the right, compared to previous versions, which opened upwards. The primary objective of the modification was to enhance the performance of the ejection system. With the old system, the time necessary to swivel the canopy forward instead of blowing it off and waiting for the seat and canopy to separate would have exceeded the minimum safe ejection height.

The MiG-21PFM was equipped with the superior KM-1 ejection seat. The KM-1 enabled the pilot to eject from zero altitude at speeds of 80 mph (130 kph) during the takeoff run and up to 745 mph (1,200 kph) in level flight.

All earlier versions of the MiG-21 were equipped with a semi-encapsulated ejection system with the integrated forward-opening canopy that remained integral with the ejection seat to shield the pilot from the blast impact until the stabilizer parachute slowed the seat's rate of descent.

Apart from the front portion, there was also a minor change in the fuselage. The air intake cone, which was significantly larger than the one on the MiG-21F-13, was composed of radio frequency transparent material with a thickness of 14 mm and covered the radar system's antenna.

Regarding the wings, they were equipped with newly designed flaps with BLC (SPS) capabilities. However, the surface of the ailerons remained unchanged. In addition, the flaps were tilted for takeoff at an angle of 25 degrees and 45 degrees for landing.

A significant change was also made to the horizontal stabilizer. There was a further increase in the chord of the tail, resulting in a combined fin and rudder area of 57.27 sq. ft. (5.32 sq. m). In addition, the new dorsal fin featured a leading edge with the same sweep angle that extended straight from the former dorsal fin's front area to a significantly larger top. This modification drastically decreased the prevalent issue of compressor surges and stalls.

As for the powerplant, the MiG-21PFM was equipped with the Tumansky R-11F2S-300 turbojet (the 'S' signifying it was for the BLC 'SPS' system). The engine delivered a maximum thrust of 13,614 lb. (6175 kg). In addition, the fuel in the internal tanks increased to 708 gallons (2,680 l), with a new metal fuel tank installed under the 'hump' fairing behind the pilot's cabin.

Additionally, the cockpit equipment was upgraded. A new type of Chromium-Nickel reconnaissance device was fitted, which not only answered queries from radar stations but also allowed the pilot to find out the nationality of nearby aircraft. In addition, an upgraded Sirena-3M Illumination warning system was fitted.

Lastly, there was a fundamental change in the armament the MiG-21PFM could carry. The fixed cannon was removed in all PFM interceptors; however, MiG added an 'RP-21M' radar system to scan for and track targets, which increased the armament load, making it possible to equip the Mig-21PFM with more missiles. As a result, the RS-2US (K-5) radar-guided missiles could be fitted on points under the wing, together with the R-3S (K-13) infrared missiles. Furthermore, an optical PKI (ASP-PF-21) shooting sight was located adjacent to the radar sight, which was used for launching the unguided missiles, bombs, and (if equipped) for firing the cannon.

An Egyptian MiG-21PFS, flying over a desert region in Egypt. (courtesy airwar.ru)

A first series MiG-21PFM 'Red 22', c/n 1112, on a snow covered Soviet airfield. (courtesy airwar.ru)

TYPE	POWERPLANT	DIMENSIONS WINGSPAN / LENGTH / WING AREA / HEIGHT	WEIGHTS Normal takeoff weight / Maximum takeoff weight	SPEED Max Speed (High altitude) / (Low altitude	SERVICE CEILING/ RANGE
MiG-21PFM Type 94	Tumansky R-11F2-300 turbojet delivering 13.117 lb. thrust	23 ft / 47.6 ft / 247.57 sq ft / 13.2 ft.	17.240 lb. / 20.106 lb.	1.386 mph / 808 mph	62.335 ft / 1.037 miles
MiG-21PFM WEAPONS		1 x 23 mm GSh-23L gun, 4 x UR Air-Air R-3S, or RS-2US missiles, one container with a twin-barreled 23 mm GSh-23 auto cannon, or PU NAR S-24, Air-Ground missile, UR Kh-66 tactical air-to-surface missiles			

MiG-21FL

This variant was an extension of the MiG-21PF interceptor manufactured exclusively for the Indian Air Force. From 1966 through 1974, the Indian national aviation company: 'Hindustan Aircraft Limited' (HAI) built 195 units of this version under license, which was designated MiG-21FL.

These MiGs were similar to their Soviet-made counterparts, but their original RP-21 radar was replaced by the R-2L radar, which was a less effective model. However, India's MiG-21FLs could employ non-Soviet made missiles, such as US, French, or other NATO-produced missiles.

To avoid disclosing the secrets of its newer Tumansky R-11F2-300 turbojet engine, Russia fitted the IAF's MiG-21FLs with the less powerful Tumansky R-11F-300 engines.

In June 1964, the first MiG-21FLs made in the Soviet Union arrived in India. The MiG-21FL first entered service with the 28th Squadron of the Indian Air Force, just like the MiG-21F-13 before.

Eight squadrons of the Indian Air Force were re-equipped with MiG-21FLs by December 1971.

That was just in the nick of time, as the Indo-Pakistani War took place that year. During the conflict, several Pakistani F-86 Sabers, F-104 Starfighters, and Chinese-built MiG-19s (designated R-6) were shot down by the IAF. Indian MiG-21s also participated in ground strike operations, bombing, and strafing Pakistani positions.

Naturally, the Soviet Union supported India in that war, while China, the United States, and the United Kingdom supported Pakistan. Thus, in the end, Russia's MiGs played a part in India's victory.

The MiG-21FL variant was manufactured under license at the 'HAI' plant in Nasik, located in the northern region of the Indian state of Maharashtra. The Tumansky engines were manufactured under license at Koraput, in the Indian state of Odisha. The Indian Air Force received its first domestically manufactured MiG-21FL in November 1966.

This is a MiG-21FL. s/n C499, at the Indian Air Force Museum in Palam, Delhi. It was assigned to 3rd Squadron. It is currently preserved at the IAF Museum at Palam AB. (Aeroprints.com)

3rd GENERATION SERIES

MiG-21R

The MiG-21R (prototype: 'E-7R') (NATO: Fishbed N) reconnaissance derivative was developed from the MiG-21PF (94 aircraft produced). It was previously labeled 'Type 03' to mislead anyone outside the Warsaw Pact. However, the official Soviet designation for this aircraft was 'Type 94R'. The letter 'R' signified that it was a reconnaissance version. The MiG-21R was also the first third-generation aircraft of the MiG-21 series. The 'R' was designed for tactical day and night photographic and electronic reconnaissance operations at low to medium altitudes.

Work on the development of the MiG-21R started in 1963 at the no. 21 Gorky plant. The plan was to create a version of the MiG-21 that could carry large unified containers that could hold photographic and electronic intelligence equipment. The prototype reconnaissance 'E-7R' versions, s/n 940724 and s/n 940735, were converted from the serial production MiG-21PFM.

Reconnaissance equipment, such as cameras and electronic gear, was to be installed on the prototypes in three separate containers: for day, night, and radio reconnaissance. That was a tall order, and MiG knew it. One large suspended tank under the MiG airframe was enough, but three were problematic. Then, MiG thought it had found the answer. It planned to fit two external beams under the MiG's airframe, with each beam holding 130 gallons (490 l) to facilitate the installation of two more suspended tanks. Be that as it may, this configuration was not the answer. The prototype reconnaissance aircraft did not pass the tests due to insufficient tactical radius of action due to the extra tanks.

Work on a prototype, based this time on the MiG-21 PFM 'Type 94'. The 'PFM' variant gained a bigger and more elaborate fairing. Furthermore, the enlarged fairing allowed for the installation of a 90-gallon (340-l) gasoline tank, increasing the total fuel supply in the internal tanks to an ample 740 gallons (2,800 l). In order to get the fuel issue right, MiG had to eliminate the no. 1 fuel tank in favor of the no.7 fuel tank, which enabled the increase in fuel capacity.

With the increase in fuel volume, tanks with a capacity of 130 gallons (490 l) were finally authorized to be suspended on the underwing pylons. In addition, four extra hardpoints were added to the wings.

Cockpit items installed on the MiG-21R were plentiful. An angle of attack sensor was fitted on the left side of the forward fuselage, and data from it was sent to the AP-155 autopilot and an indicator on the pilot's dashboard. The air pressure receiver, which had previously been situated in the center of the aircraft, obstructing the pilot's view, was relocated to the right, improving visibility. A backup air pressure receiver was fitted on the right side of the fuselage in front of the cockpit. In addition, an SPO-3 electromagnetic exposure warning receiver and a preserved RAS TsD-30 radar were also among the MiG-21R's onboard equipment.

A container with special equipment was installed under the fuselage of the MiG-21R on a unique streamlined holder. The container included the following items;

- 'D' for daytime photo reconnaissance - cameras for perspective shooting 2 x AFA-39, cameras for planned shooting 4 x AFA-39, slit camera AFA-5;
- 'N' for night photo-reconnaissance - camera UAFA-47, lighting photo cartridges 188 pcs.
- 'R' for electronic intelligence equipment 'Romb-4A' and 'Romb-4B', camera AFA-39 for control;
- A for active Interference System 'Siren'
- A for air sampling equipment;
- E for equipment for relaying audio information in the VHF band.

With the large number of changes made to the 'PFM' variant, the aircraft was designated MiG-21R. It was powered by an R-11 F2S-300 with 123,670 lb. (6,200 kg) thrust.

The jet was produced in the Gorky aircraft plant between 1965 and 1971. Interestingly, two Warsaw Pact Air Forces continued to deploy this type into the 1990s and beyond 2000. Between 1969 and 1972, the Czechoslovakian Air Force received the MiG-21R reconnaissance version. It remained in use when the nation was split into the Czech Republic and Slovakia in 1992. Following that, several of the MiG-21Rs were transferred to Slovakia.

In Poland, the first MiG-21R was delivered to the Polish AF in September 1968, and the final one was supplied in 1972. Twelve Mig-21Rs were provided to the Polish Air Force and served until 2002.

Incidentally, MiG supplied Egypt with the MiG-21R. It was given a local designation; MiG-21RF (photo-apparatus). It was equipped with 3 A-39 cameras permanently installed behind the front landing gear bay.

This is one of the 'E-7R' prototype reconnaissance jets, which took part in the suspended tank tests. The 'reconnaissance' suspended tank can be seen beneath the central area of the aircrafts fuselage. (courtesy airwar.ru)

This MiG-21R is one of the first production aircraft. (courtesy airwar.ru)

This MiG-21R was delivered to the Romanian AF in 1972. It served with unit EdeR 31 till 1998, when it was wfu, and stored at Craiova AB. It is preserved at Pres. Muzeul Militar National. (Jim van de Burgt NL)

An Egyptian MiG-21RF. Note the recon capsule under the fuselage. (LeCaire)

This Mig-21R served in the Polish AF for 25 years. It was delivered to the PAF in November 1969, and it was retired in 1994. It served with two regiments; 1969-1982 with 21 PLRT (Pułk Lotnictwa Rozpoznania Taktycznego / Tactical Reconnaissance Aviation Regiment), and from 1982 to 1994 with 32 PLRT. It is now preserved at Muzeum Bitwy nad Bzur. (Wistula)

This is another Polish AF MiG-21R, s/n 1125, c/n 94R021125. It was delivered to the Polish AF in January 1972, and served with 32 PLRT from 1972 to 1996, when it was WFU. Currently, it is preserved at the Polish Aviation Museum in Krakow. (Alan Wilson)

TYPE	POWERPLANT	DIMENSIONS WINGSPAN / LENGTH / WING AREA / HEIGHT	WEIGHTS Empty / Maximum takeoff weight	SPEED Max Speed	SERVICE CEILING/ RANGE
MiG-21R Type 94R	Tumansky R-11F2S-300 turbojet with 8,708 lb. dry, and 13,613 lb. with a.b.	23.47 ft / 46.2 ft / 247.57 sq ft / 13.5 ft.	11,795 lb. / 17,857 lb.	1,056 mph	49,540 ft/ 994 miles
MiG-21R WEAPONS	Two R-3S melee missiles, or two UB-16-57U (UB-32) NUR type S-24 launchers, or 2 bombs.				

MiG-21S

The first MiG fighter of the 3rd generation was the MiG-21S. Flight tests of the MiG-21S 'Type 95' (Fishbed-H) began in 1964, and in 1965 The Gorky factory produced the first 25 aircraft.

The Mig-21S was fitted with the RP-21 'Sapfir 21' (Sapphire 21) (NATO: Spin Scan) radar. The 'Sapfir 21' radar was also installed on the MiG-21PM. However, unlike the MiG-21PF, the MiG-21S had a larger-capacity fuel tank in the enlarged fairing behind the cockpit. In addition to the 'Sapfir 21' radar, the 'S' variant also had two more hardpoints under the wings inherited from the MiG-21R.

In contrast to the MiG-21R, however, the MiG-21S variant did not have provisions for suspending reconnaissance and electronic warfare canisters beneath its airframe. Instead, a more lethal GP-9 gun container was fitted beneath the jet's fuselage.

Although this appeared to be a good thing, the gun canister generated moee drag for the MiG-21S, hence it was rarely deployed on fighters.

Because the MiG-21S was largely utilized as a fighter-interceptor, it rarely carried the GP-9 cannon. The GP-9 weapon container was merely an interim measure until the MiG-21S' fuselage could be outfitted with a permanent, semi-recessed, double-barreled GSh-23L cannon. On the plus side, the 'S' version was equipped with four K-13 missiles.

The MiG-21S was also equipped with the advanced RP-22S radar (NATO: 'Jay Bird'), the ASP-PF sight, and a 13,611 lb. (3,248 kg) thrust Tumansky R-11F2S-300 turbojet engine. The MiG-21S version was mass-produced between 1965 and 1968 for domestic (USSR) use only. Not one MiG-21S was ever exported.

These two MiG-21S jets were equipped with the 'Sapfir 21' (Sapphire-21) radar system, and were fitted with larger fuel tanks. The 'S' version also carried more weaponry, thanks to the two extra hardpoints under the wings. Note the two rocket cannisters on the aircraft in the bottom photo. (courtesy airwar.ru)

MiG-21SM

The MiG-21S was quickly superseded by the MiG-21SM 'Type 15' (NATO: Fishbed H). The 'SM' variant was what the 'S' version should have been. Basically, the MiG-21S was a heavier aircraft than its predecessor, the MiG-21R. Still, it was powered by the less formidable R11F2S-300 turbojet that was fitted on, the lighter MiG-21R.

The MiG-21SM fighter interceptor fixed the problem by being equipped with the more potent Tumansky R-13-300 engine supplying 14,550 lb. (6.600 kg) thrust. The two-stage turbojet engine featured a five-stage high-pressure compressor, a combustion chamber design that made restarting the engine at high altitudes easier. It also had a new afterburner and made greater use of titanium components.

A GSh-23L gun with 200 rounds of ammunition was fitted beneath the fuselage, in front of the centerline pylon, on the MiG-21SM. In addition, the ASP-PF optical sight was replaced with the AP-PFD sight, which had a built-in cannon. On the four APU-13 launch terminals, various combinations of R-3S, R-3R, UB-16-57UM, or UB-32 blocks, bombs, and incendiary tanks could be suspended. While on the APU-7 launch terminals, older RS-2US missiles could be used instead of the R-3R missiles.

In ground attack mode, the MiG-21SM could carry two 1,100 lb. (500-kg) bombs and a 130-gallon (490 l) tank, or eight 220 lb. (100 kg) bombs and a 211-gallon (800 l) tank.

The 'SM' could also be equipped with an A-39 camera. The camera was placed behind the recess of the front landing gear.

With the extra weaponry and other equipment on the aircraft, capacity increased to 2,866 lb. (1300 kg). That led to a reduction in fuel capacity to 709 gallons (2,685 l). But, that was compensated by enlarging the additional tank under the airframe, which increased the fuel capacity by 211 gallons (800 l).

The avionics of the MiG-21S were cutting-edge. They included the 'Sapfir 21' radar, which used 'vacuum tube' technology and was easy to use and repair, the 'SPO-Yu' radiation warning system, an 'ASP-PFD' optical sight, and an 'S-21' radio sight. The 'Flight-OI' Flight and Navigation Complex (PNK) was also installed aboard the jet, along with the 'SVU-23ESN' automated control system, the 'RSBSN-5S' short-range navigation and landing system, and the 'Pion-N' antenna-feeder system. It also had the 'Lazur' noise-proof communication line, which allowed it to communicate with the Vozdukh-1 ground-based automated control system.

The MiG-21SM, just like the MiG-21S, was produced only for the Soviet Air Force.

This is a MiG-21SM equipped with UR R-3S missiles, and PTB-490 cannon. (courtesy airwar.ru)

A MiG-21SM with a UB-16-57UM rocket pod, seen on the right wing hardpoint. (courtesy airwar.ru)

This is the prototype MiG-21SM. (courtesy airwar.ru)

This MiG-21SM 'Red 70' is preserved at Central Armed Forces Museum (CMOTAF) in Moscow. (Alan Wilson)

MiG-21M (E-7M)

The MiG-21SM was the basis for the MiG-21M, initially 'Type 94', revised to 'Type 96' (NATO: Fishbed-J. It was the export version of the MiG-21SM variant. However, the two fighters were not comparable. The MiG-21M was a watered-down variant of the MiG-21SM. At the same time, the latter was powered by the more powerful R-13-300 engine, while the MiG-21M was equipped with the less formidable R-11F2S-300 turbojet. In addition, instead of the advanced RP-22 'Sapphire 21' system, an older radar, the RP-21 MA, -RP-21MA, coupled with the ASP-PFD automatic sight, was installed on the 'M' variant.

The installation of the 'RP-21MA' was contentious. The pilots, noting the high maneuverability of the fighter and ease of its control, complained about the weakness of weapons, including the lack of built-in guns and the primitive sight. As a result, the Gryazev-Shipunov GSh-23L twin-barrel 23 mm autocannon with 200 rounds of ammunition was mounted on a carriage and placed beneath the fuselage. However, that meant the MiG had to remove one fuel tank, to accommodate the cannon. However, the gun could be pushed down during maintenance work, providing convenient access to the jet's ordnance.

There were some improvements. The internal fuel tank was replaced with a bigger one, comparable to the MiG-21S' tank, to compensate for the reduced fuel capacity. To further increase fuel load and range, three 132 gallon (500 l) tanks or two 132-gallon (500 l) and 211-gallon (800 l) tanks could be suspended under the wings.

The wings of the MiG-21M had four hardpoints beneath them, allowing for more ammunition. In addition to the UR R-3C (K-13) missiles, the MiG-21M could now be armed with the previously prohibited for export 'RS-2-US' missiles.

It could also carry up to four 'UR K-13M', 'K-13' missiles, NAR rockets in UB-16 or UB-32 blocks, several types of free-fall bombs weighing up to 220 lb. (100 kg), or incendiary tanks. The 'M' variant's maximum combat load weight was 2,646 lb. (1200 kg).

From 1968 to 1971, the Moscow machine-building facility 'Znamya Truda' produced the MiG-21M, which was delivered to numerous nations throughout the world. Subsequently, India obtained the license to manufacture the fighter, and the first copy of the MiG-21M produced in India was delivered to the Indian Air Force on February 14, 1973.

Images by Newresid stephanelhernault@yahoo.fr

A Bulgarian AF MiG-21M, the export variant for Warsaw Pact, and other friendly countries. Delivered in August 1980, it served with 2/19 IAE in 1970, then in 1984 with 1/21 IAF, before WFU in 1990. It is preserved at the Krumovo AB museum. (Richard Vandervord)

MiG-21M of the Polish AF, c/n 961904 s/n 1904. Delivered to the PAF in March 1970, and it served with the following fighter regiments: 1970-1988 9 PLM, 1989-1993 2 PLM, 1993-1996 1 PLM, 1996-1998 41 PLM, 1998-1999 11 PLM, 1999-2000 10 PLM. It has been preserved at Savigny les Beaune, France since January 2003. (Richard Vandervord)

MiG-21PD (E-7PD)

A derivative of the Mig-21 was the E-7PD 'Type 92', MiG-21PD (NATO: Fishbed-G). It was an experimental aircraft with STOL capabilities. The E-7PD variant (product 23-31) was derived from the MiG-21PFS. The jet was primarily designed for testing the STOL lift engines of the MiG-23PD (23-01) fighter.

The MiG-21PD (E-7D) was equipped with a Tumansky R-13F-300 main engine delivering 15,100 lb. (6,850 kg), and two RD-36-35 STOL engines were positioned behind the cockpit in the aircraft's center of gravity. The MiG-21-PD's role was to test the two STOL 'lift' engines. In order to do so, the MiG-21D's fuselage was lengthened by 2.95 ft. (900 mm). In addition, air was supplied to the STOL engines through air supply ducts that opened during takeoff and landing. The two STOL lift turbojet engines each had a thrust of 5,180 lb. (2,350 kg). Air to the two engines was supplied through ducts that opened during takeoff and landing.

Due to the R-13F-300 air intakes going through the main landing gear bay area, the E-7PD prototype's landing gear could not be retracted. Due to this, the jet's maximum speed was severely restricted. In addition, the installation of lift engines in the center part of the aircraft necessitated a significant reduction in the internal fuel supply. However, because the prototype E-7PD's propulsion system consisted of three engines, this model was a gas guzzler. Therefore, the MiG-21PD's total flying endurance did not exceed 17 minutes.

P.M. Ostapenko was the test pilot who made the MiG-21PD's maiden flight on June 16, 1966 and participated in the air demonstrations at the Domodedovo Aviation Parade in Moscow in June 1967.

The E-7PD's takeoff and landing procedures were not straightforward. At ground level, the flow of exhaust gases from the lift engines dispersed in all directions and reduced the wings' lift. In addition, the effect of the STOL engines on the total buoyancy, stability, and controllability was highly dependent on the flying altitude and speed. During takeoff, the inclusion of the E-7PD's lift engines caused only a modest and temporary change in longitudinal stability. Furthermore, their activation during the landing maneuver resulted in a sudden loss of altitude and speed. Before landing, it was therefore necessary to increase the thrust of the main propulsion unit to the maximum, or ignite the afterburner chamber. In the end, the E-7PD test program demonstrated that STOL combat aircraft with lift engines were quite impracticable.

The MiG-21PD (E-7D) 'Fishbed G'. Note the opened air supply duct behind the cockpit. (courtesy airwar.ru)

The prototype E-7PD. (courtesy airwar.ru)

The streamlined shape of the MiG-21PD with the 2.5 foot extended airframe. (courtesy airwar.ru)

The MiG-21I 'Analog'

The engineers at the Tupolev Design Bureau wanted data on the actual flying behavior of an aircraft with an ogival wing when developing the Tu-144 supersonic passenger aircraft.

A small experimental analog aircraft, the Mig-21I, was designed for this purpose. It was equipped with the powerful Tumansky, R-13F-300 engine with 15,100 lb. thrust (6,850 kg). The first flight took place on April 18, 1968, with test pilot O.V.Gudkov at the controls.

Apart from determining the flying characteristics of the ogival winged plane, the 'Analog' jet was also used to test the flight control system. As a result, it was the ideal 'simulator' for the Tu-144 crew. The first three test pilots completed a distinct flight program on the 'Analog' aircraft to develop flying methods. Subsequently, the 'Analog' accompanied the Tu-144 on its inaugural flight on December 31, 1968.

The airframe of the MiG-21S fighter and its vertical tail and landing gear were used as a basis to construct the 'Analog' aircraft. The Mig-21I had a variable 78-degree sweepback on 30% of the wings and a 55-degree sweepback on the remaining 70%. Each half-wing was equipped with four sections of 'tailerons' that ran the length of the wing. The main landing gear struts were retracted into the wing and fuselage. In addition, there were four sections with braking flaps and a ventral fin beneath the fuselage. The aircraft also included control, measuring, and telemetry systems.

Between April 18, 1968, and December 1968, a total of 70 test flights were conducted, 64 of which were completed on the 'Analog' jet at subsonic and supersonic speeds. The first set of pilots then continued their flight training on the Tu-144. After that, the rest of the pilots followed suit.

By July 1969, another 142 test flights were completed on the experimental MiG-21I 'Analog' fighter. During those tests, the pilots reached an altitude of 62,335 ft. (19,000 m) with speeds ranging from 132 mph (212 kph) to Mach 2.06.

On July 28, 1970, after the primary test program was completed, one of the test pilots crashed the prototype Mig-21I during aerobatic tests. That set the Tu-144 program back a few months until the second 'Analog' type became ready at the end of 1969. Training returned to normal by the end of January 1970.

The early flights of the 'Analog' jet were primarily geared towards figuring out methods of controlling tailless aircraft. The second Mig-21I, on the other hand, was tasked with investigating the change in the shape of the wing inflows. To accomplish that task, film cameras were put in the aircraft's streamlined fuselage and on the top of the fin. A laser imaging system of the wing flow was also installed on the plane. In addition, full and static pressure sensors were placed on its top and lower surfaces. Flights of the MiG-21I were conducted at altitudes of up to 7.45 miles (12 km) and Mach speeds of up to 2.06. They flew with varying alignment values to replicate the behavior of the future supersonic airliner, which were adjusted by attaching weights ranging from 610 lb. (277kg) to 639 lb. (290 kg) in the nose or tail regions of the fuselage.

Tupolev's flying experiments on the MiG-21I allowed the firm to investigate nearly all of the tailless aircraft's flight dynamics and manageability characteristics. Furthermore, it analyzed the 'Analog's' aerobatic capabilities over various angles of attack and Mach numbers.

The MiG-21I 'Analog' proved to be an uncomplicated and remarkable aircraft to fly. The pilots had a great time flying it. It was exceptionally maneuverable due to its larger wing area. However, it accelerated slower than a standard MiG-21 fighter. Finally, after the testing of the Tu-144 pilots had finished, the second MiG-21I was transported to Monino Air Force Museum in Moscow.

Three different views of the MiG-21I 'Analog' supersonic trainer. For the test pilots, it was a joy to fly the 'Analog' (courtesy airwar.ru)

E-8/1 /2

Another MiG-21 experimental was the E-8, which was heavily modified from the MiG-21PF. This variant had a very different shape from a typical MiG-21. The E-8 only shared a double delta wing, main landing gear, a steering system, and a hydraulic system with the 'PF' version, and it looked more like the Eurofighter Typhoon than a MiG-21. Only two E-8s were built.

The E-8's primary objective was to shoot down enemy fighters with missiles in all weather conditions and at any time of day or night. While its secondary objective was to destroy ground targets.

During the design process in 1960, the E-8 was designated MiG-23 because it was created for the S-23 weapons system. The S-23 complex comprised the 'Sapphire-1' radar with pulsed radiation (later, in the second stage, Sapphire-2 with quasi-continuous radiation), two K-23 missiles, the ASP-PF sight, and the 'Spektr' infrared direction finder. The S-23 system would enable the E-8 to gain air superiority.

However, that was not to be. Due to the S-23 armament system's unavailability in 1961, the E-8 aircraft was redesignated MiG-21M. That was because it was outfitted with the TsD-30TP radar, the Samotsvet infrared sight, and two K-13 infrared missiles.

The E-8 had the more powerful R-21F-300 engine since it was heavier than the MiG-21PF. Because this engine was larger, it required an expansion of the fuselage's whole center and rear section to accommodate it. Compared to the R11F-300's 2.53 ft. (772 mm) diameter, the R21F-300's girth was 845 mm. In addition, the thickness of the afterburner shroud increased from 902 mm to 987 mm, and the dry weight increased from 2,570 lb. (1,165 kg) to 2,755 lb. (1,250 kg). As a result of the radical changes in the back of the fuselage, the front section also required a complete redesign of the nose landing gear structure.

Furthermore, the OKB MiG designers sought to prevent problems caused by foreign object penetration into the power unit's air supply during taxiing. Thus, the engine had to be shifted further back. As a result, the E-8's front section was tilted slightly upwards.

The E-8, like the E-6T/3 before it, had canards in the front of the airframe, just in front of the cockpit. The canards moved freely with the airflow at subsonic and low supersonic speeds, but at speeds more than Mach 1.9, they were locked in the neutral position. The installation of the canards brought about an increase in the value of aerodynamic smoothness and agility.

Being experimental aircraft, the nose cones of the E-8/1 and E-8/2 contained blocks of test equipment rather than radar. For this reason, their cones were made of metal.

On March 5, 1962, the factory introduced the first E-8/1 prototype, also known as 'Red 81.' On April 17, 1962, test pilot G.K. Mosolov took it for its first flight. Then, on July 13, 1962, the second experimental, the E-8/2, 'Red 82', made its maiden flight with test pilot A.V. Fedotov at the controls.

Many tests were conducted with the two E-8s. The E-8/1 was tasked with fine-tuning the powerplant and ensuring the stable operation of the R-21F-300 engine. It carried out 25 flights.

Then, tragedy struck. On September 11, 1962, while flying at supersonic speed, the E-8/1's engine exploded, damaging the control system. Mosolov was at the controls, and he had two choices, either crash with the plane or eject from it at high speed. Mosolov chose the latter. Just a few seconds after ejecting, another explosion ripped the plane apart. However, fragments from the exploded aircraft hit Mosolov, who lost consciousness. The parachute landed in a forest, where the test pilot was found and transported by helicopter to the hospital.

Mosolov's condition was severe, and clinical death occurred twice. Employees of MiG's OKB joined in to save their co-worker's life.

They even built a movable bed to change Mosolov's position without injury. After many months of continuous treatment, Mosolov recovered. Regrettably, he never flew again.

Regarding the E-8/2, it completed at least thirteen flights. However, the loss of the first prototype and engine problems with the E-8/2 resulted in the termination of the experiments and the project as a whole.

The photos above and below, are of the E-8/1, 'Red 81'. It was the aircraft that exploded in mid-air with the unfortunate Mosolov in the cockpit. The top three photos on the right are of the second E-8/2, 'Red 82' experimental. (courtesy airwar.ru)

MiG-21MF

The MiG-21MF 'Type 96F' (NATO: Fishbed J) was created based on the MiG-21M. The letter M stood for 'Modernizirovannyy' (Modernized), and the F was 'Forsirovannyy' or (Uprated engine).

The main difference between the MiG-21M and the MiG-21MF was that the 'MF' was equipped with the more powerful Tumansky R-13-300 turbojet engine offering 14,308 lb. (6,490 kg) thrust, compared to the MiG-21M's R-11F2S-300 engine with 13,613 lb. (6,175 kg) thrust. Nevertheless, it was the same engine as the MiG-21SM. Furthermore, the 'MF' version also had a more powerful radar, the Rp-22SMA 'Sapphire-21', in place of the type Rp-21MA.

The MiG-21MF's internal tanks could hold 700 gallons (2,650 l). In addition, a 210-gallon (800 l) fuel tank could be suspended from the central ventral pylon, and another two 490-liter tanks could be slung on the wing's outer pylons.

Aside from the fact that the USSR Air Force exclusively used the MiG-21SM, the MiG-21MF differed from the 'SM' variant solely by the installation of less sophisticated avionics equipment in place of the 'secretive' avionics systems that were only to be used by the MiG-21SM.

Moreover, the airframe of the MiG-21MF featured many design enhancements intended to facilitate flights. Thus, titanium was employed in place of steel for various power structures. Additionally, the MiG-21MFs were fitted with the 'SARPP-12' system for fixing flight parameters. This system was created to record boundaries under normal and emergency conditions and use this data for flight crew training and accident investigations.

Regarding weaponry, the MiG-21MF was fitted with air-to-air missiles, such as the K-13A infrared and R-60 missiles, designed for close-in aerial combat. Two triggers were placed on the control stick; one was for launching the missiles, and the other was for firing the GSh-23L cannon.

The MiG-21MF was also a formidable ground assault aircraft. It could strike ground targets since its suspended armament arsenal contained bombs weighing up to 1,100 lb. (500 kg) and NAR in UB-16 or UB-32 blocks containing unguided rockets.

The MiG-21MF was officially rolled out in 1969. Soon after, the MiG-21MF and the MiG-21M began to leave the Moscow 'Znamya Truda' factory's production line in a steady stream. By the end of 1969, 96 MiG-21MFs had been produced, and they were all quickly expedited to Egypt.

Then, from 1970 to 1974, the MiG-21MF was serially manufactured in the Moscow Aviation Plant 'Znamya Truda' before being transferred to the Gorky plant in 1975.

MiG-21MF fighters were widely used in many conflicts in the late Cold War years, ranging from Afghanistan to the Iran-Iraq war of 1980-1988 and during the first Desert Storm in Iraq. In addition, the MiG-21MF was sold to dozens of countries worldwide that did not use it in conflicts.

	MiG-21MF 'Type 96F'
Powerplant	**R13-300 14308. lb. thrust**
Wingspan	**23.45 ft.**
Length	**48.8 ft.**
Height	**14.45 ft.**
Wing Area	**247.57 sq. ft.**
Empty Weight	**11,795 lb.**
Takeoff Weight	**20,500 lb.**
Max. Speed	**1,367 mph (High altitude) 808 mph (Low altitude)**
Range	**882 miles**
Service ceiling	**59,710 ft.**

This MiG-21MF, c/n 969309, s/n 9309, was delivered to the Hungarian AF in May 1974. It served with the 59 HVE for the whole of its service life. In 1997 it was wfu. It is currently on display at the Szolnok museum. (Richard Vandervord)

This East German MiG-21MF was received by the AF in May 1975. It served with JG-9, then with JG-3. After the unification of Germany, it was absorbed into the newly integrated German AF. In October 1990, it was wfu. It is currently preserved at the Motortechnica Museum in Bad Oeyenhausen, but, its falsely marked as '22+28', when it should be **'23+48'** (Rob Schleiffert)

This MiG-21MF, c/n 967711, s/n 7711, served with nine different units. It was delivered to the Czechoslovakian AF in June 1973. Its first stop was with 6 SLP in 1973, then 1988, 9 SLP, 1993, 11 SLP, 1993, 82 IFS, 1994, 4 IFS. In 1994 it was with the Stressova Letka Fight Demonstration Unit unit, at Caslav AB. In 1994, with 11 SLP, later in 1994 with 1 SLP, and in 1995 with LZO-LZS no. 1, at Line AB. Then tragedy struck. On June 8, 1998 The Mig was in a mid-air collision with MiG-21UM '3756' over the base at Ceske Budejovice. Thankfully, Major Lubos Kubat ejected safely. The MiG, however, was destroyed. (Richard Vandervord)

This stylish Czechoslovakian MiG-21MF, c/n 967701, s/n 7701, was delivered in June 1973, it was assigned to Fighter Regiment 6 SLP. In 1988 in joined 9 SLP, then 11 SLP in 1993. In 1994, it changed track, it became a flight demonstration jet. It joined the Stressova Letka demo unit, at Caslav AB. In 1998 it moved to 4 zTL (4th Tactical Air Force Base), which was when this photo was taken. In 1999 it joined 41 SLT, and in March 2002, it was stored at Ceske Budejovice AB. (Simon Butler)

4th GENERATION SERIES

MiG-21MT

The MiG-21MT 'Type 96T' was a distinctive variant derived from the MiG-21M interceptor with increased internal fuel tank capacity due to a more prominent dorsal fairing behind the cockpit. The fuel supply in the MiG-21MT's internal tanks expanded to 860 gallons (3,250 l). The 'MT' was intended to be the export version of the MiG-21SMT. The MiG-21MT's armament and radar were comparable to those of the MiG-21MF. The Moscow-based 'Znamya Truda' plant manufactured 15 MiG-21MTs in 1971, but none were exported. Instead, five were transferred to the USSR Air Force test-pilot school, and one was used as a test and research aircraft by the MiG OKU.

MiG-21SMT

The MiG-21 SMT 'Type 50' (NATO: Fishbed K) was the first variant of the fourth generation.

It was a cross between the MiG-21SM and the MiG-21MT. It had the same internal tank capacity as the 'MT,' but its tanks were placed differently in the dorsal spine behind the cockpit. In their constant desire to increase the MiG-21's internal fuel capacity, the MiG designers decided to expand the size of the dorsal fairing even more. Between 1969 and 1970, they increased it to a level that made the plane look awkwardly un-aerodynamic.

The massively extended spine was designed to contain a row of three saddle tanks with a combined capacity of 237 gallons (900 l), bringing the total internal fuel capacity to 859 gallons (3,250 l). As a result, flight performance inevitably decreased while the takeoff run increased. However, the OKB test pilots believed the reduced in-flight stability margin to be unproblematic. Nonetheless, it was subsequently determined that the massive bulk of fuel high above the axis of rotation and the center of gravity's rearward migration would make the MiG-21SMT unsafe for inexperienced pilots. As a result, OKB chose to take out the last tank at the back of the dorsal spine. This reduced fuel capacity by 79 gallons, (300 l) to 779 gallons (2,950 l).

The Soviet combat pilots who flew the MiG-21 SMT did not like it at all. Consequently, serial production of the MiG-21SMT did not last long. The Mig-21SMT was only produced between 1971 and 1972, and, even then, in relatively small quantities. The MiG-21SMTs entered service exclusively with the Soviet Air Force, as they were not for export. During their service, they were used, not only as fighters, but also as carriers of tactical nuclear weapons.

This MiG-21SMT was left behind in Altenburg Germany, by the Russians, after the collapse of the Soviet Union. (Jerry Gunner)

MiG-21SMT code 71, c/n 961804, is preserved at the Samara State Aerospace University Museum, Russia. (Alan Wilson)

This MiG-21SMT is the same one as in the previous page. Note the expanded dorsal spine on the SMT. (User:WikiABG)

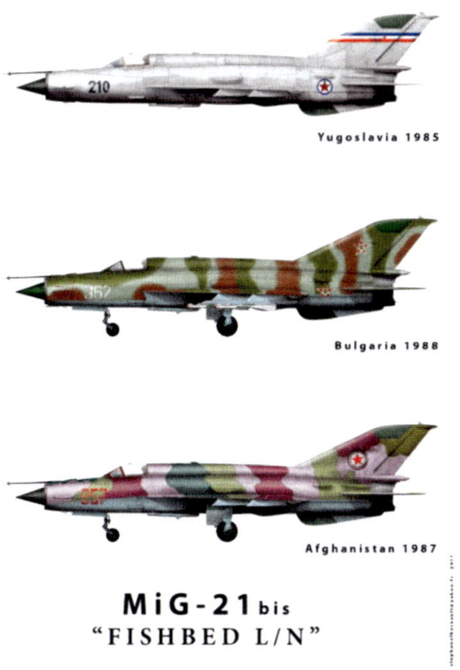

Images by Newresid stephanelhernault@yahoo.fr

MiG-21bis

The last of a great breed was the MiG-21bis (E-7bis), where 'bis' is a Latin term meaning 'once again' in US English. So it would be the equivalent to 'II,' as in F-4 Phantom II, or in British English to F-4 Phantom MK II). The MiG-21bis was labeled 'Type 75' (NATO: Fishbed-L / N).

The MiG-21bis (Fishbed L) was designed based on the airframe of the MiG-21MF. However, the 'bis' airframe differed significantly from the 'MF.' For example, most metal parts were replaced with titanium, which has heat resistance capabilities that made the Mig-21bis lighter, more maneuverable, and more durable.

The MiG-21bis powerplant was the formidable Tumansky R-25-300 engine, rated at 9,020 lb. (4,100 kg). At take-off, its thrust increased to 15,653 lb. (7,100 kg). Moreover, with a particular afterburning procedure at speeds above Mach 1, the MiG could receive 21,825 lb. thrust (9,900 kg) for three minutes at altitudes up to 13,120 ft. (4,000 m). Furthermore, MiG enlarged the jet's dorsal spine allowing for a larger internal fuel tank on the 'bis.' That enabled the fuel capacity to be raised by almost 60 gallons (230 l) to 762 gallons (2,885 l). The upgraded engine increased the MiG-21bis' climbing speed to 738 feet per second (225 meters per second) from 426 feet per second (130 meters per second) in the previous MiG-21 SM variant. In addition, the acceleration time from 373 mph (600 kph) to 684 mph (1,100 kph) decreased from 28 sec. to 19 sec.

The MiG-21 bis 'L' was armed with an internal GSh-23L cannon, which allowed it to carry the R-55 'AA-1 'Alkali' and R-60M (AA-8 'Aphid' missiles on underwing pods. The missiles had double the range of the older R-3S 'Advanced Atoll' missiles. In addition, the R-55 and R-60M dogfight missiles could be launched with a high load factor.

The MiG-21bis (Fishbed L/N) was manufactured in two variants: the (VVS) Soviet Air Force variant (Fishbed N) and the PVO USSR Air Defense variant (Fishbed L). The Air Defense MiG-21bis 'L' was fitted with the Lazur-M type navigation datalink, which guided the aircraft to air targets using automated ground guidance stations of the 'Vozdukh-1' system.

The VVS MiG-21'bis' 'Fishbed N' variant was equipped with the 'Sistema Avtomaticheskovo Upravleniya' (Polyot ILS system) SAU - 23JeSN, and the 'RSBN-2S' 'Radio Engineering System' for close navigation, which was later upgraded to the 'RSBN-5S' model. Both installations allowed for more accurate navigation and an instrument (automatic) approach to landing in difficult weather conditions. At the same time, the Lazur-M equipment was uninstalled from the MiG-21bis 'N' versions.

The MiG-21bis 'N' was easily identified by the distinctive RSBN antennas. They were positioned under the air intake and on top of the vertical stabilizer. Unofficially, these 'bis' were known as the MiG-21 bis SAU.

The MiG-21bis proved to be an outstanding fighter at low and medium altitudes, although it had several flaws. In particular, its radar was ineffective, as its target detection range did not exceed 25 miles (40 km). Moreover, the radio sight was incapable of tracking air targets against the background of the ground. Furthermore, despite an increase in fuel capacity from the installation of larger internal tanks and a decrease in engine fuel consumption, the MiG-21Bis' range and flying duration were relatively inadequate.

The MiG-21bis fighter was only mass-produced at the Gorky no. 21 plant from 1972 until October 1974. The 'Type 75A' Mig-21bis was available to the Warsaw Pact countries, whereas the 'Type 75B' was available to the rest of the world. In all, 2,030 MiG-21 bis were produced at the Gorky plant. In India, licensed manufacturing of the MiG-21bis began in 1980 and ceased in 1987. During that time, the HAL Nasik Division manufactured 200 MiG-21bis, with 30 to 50 aircraft produced per year.

The prototype MiG-21bis with two UR R-3S 'AA-2', (NATO: 'Atoll B') missiles, and four UR R-60 'AA-8' (NATO: 'Aphid') missiles. (courtesy airwar.ru)

Another shot of the MiG-21bis prototype. (courtesy airwar.ru)

Hungarian AF MiG-21bis (SAU), c/n 75061904, s/n 1904. Delivered in May, 1978. This is the 'Fishbed 'N' with the distinctive RSBN antenna. (Note the antenna under the air intake). It was assigned to unit 31 HRE in 1993, and was wfu at Papa AB in 1996. It is currently preserved at the Szolnok AB Museum. (Richard Vandervord)

This extremely clean and shiny MiG-21bis belongs to the Nth. Korean AF. C/n 75014205, s/n 42. The North Koreans purchased it from the Kazakhstan AF in 1999. It is supposedly still in active service. (Richard Vandervord)

East German AF MiG-21bis 'L' with LAZUR. C/n 75033205 s/n 874. Delivered in October 1975, and served with units JG9 and JG8. It was absorbed into the unified German AF and given code 24+29. WFU October 1990. Currently at Beauvechain AB 1st Wing Museum, Belgium. (Richard Vandervord)

A MiG-21bis in Finnish AF colors. C/n N75084608, s/n MG-138. It was delivered in September 1980, and served with 31 Squadron throughout its service life. It is currently at the Finnish Air Force Museum, at Tikkakoski, which is next to Jyväskylä Airport. (Richard Vandervord)

This is MiG-21bis, code no: 48, c/n 75021148. It saw service with the Soviet PVO (Air Defence Force) with the 'Lazur GCI' system (Fishbed-L) and the Soviet Air Force 'Polyot ILS' system (Fishbed-N). It is currently at the VVS museum; Monino, Russia. (Alan Wilson)

This is a Polish AF MiG-21bis, c/n 75089204/04, s/n 9204. Delivered in July 1981, this aircraft was first assigned to 26 PLM. In 1989 it was allocated to 9 PLM, then 9 ELT. Its last placement was with 3 ELT. WFU on December 31, 2003. It is currently preserved at Krakow-Rakowice museum (Polish Aviation Museum). (Alan Wilson)

TYPE	POWERPLANT	DIMENSIONS WINGSPAN / LENGTH / HEIGHT	WEIGHTS Empty / Maximum takeoff weight	SPEED Max Speed / Landing speed	SERVICE CEILING/ RANGE
MiG-21bis Type 75	Tumansky R-25-300 turbojet with 15.344 lb. dry. and 21.407 lb. with a.b.	23.47 ft / 44.2 ft / 247.57 sq ft / 13.5 ft.	11.785 lb. / 17.549 lb.	1.386 mph / 168 mph	62.336 ft/ 1.150 miles
MiG-21bis WEAPONS		1 x GSh-23L 23mm cannon (200 rounds). Maximum combat load was up to 1300 kg. Four of the following ordnance could be carried on underwing hard points: UR K-13M, RS-2US, R-ZS, R-ZR, R-60, R-60M. Additionally, NAR caliber 57 and 240 mm, and free-fall bombs of various types caliber up to 500 kg could also be carried.			

MiG-21 Trainers

The pilots of the MiG-21 aircraft were regarded as national heroes by the Soviet people. Their MiG-21 fighter jets were seen as the Soviet Union's defenders against the wicked West. During the Cold War, the citizens of the USSR were familiar with both the MiG fighter jets and their 'Ace' pilots.

However, three MiG planes were relatively unknown to most Soviet Union residents. Perhaps it was because they were 'boring trainers' rather than fully armed fighters. The MiG-21U, MiG-21US, and MiG-21UM were those three trainers. They, and the 'flight instructors' who taught the cadets to become good fighter pilots, were the unsung heroes of the Soviet Air Force.

MiG-21U (E-6U)

In the early 1960s, even though they had to fly the new supersonic 1960s jet fighters when they graduated, Soviet Air Force cadets were still taught to fly on obsolete subsonic MiG-15UTI jet trainers. To address this issue, the Air Force requested a new compact, two-seat, tandem jet trainer based on the supersonic MiG-21F/F-13. The request was approved, and on November 11, 1959, OKB MiG was tasked with developing the two-seater combat trainer.

The new supersonic two-seater, powered by the R-11F-300, was first designated 'E-6U' and, once in official service, as the MiG-21U ('U stood for 'Uchebnyy' or 'Training') 'Type 66', (NATO: Mongol A). It was unofficially nicknamed 'Sparka' (Twin) by its pilots.

Soon after its introduction, it became the primary aircraft of Soviet fighter aviation schools. The E-6U, 'Red 61', took to the skies for the first time on October 17, 1960, with test pilot P.M. Ostapyenko in command. All was well until April 1961. That month, state tests were halted, and the E-6U prototype was returned to OKB MiG to address various flaws, the most significant of which was longitudinal stability.

In July 1961, the second 'E-6U' prototype, 'Red 62', was the first to be equipped with an extra 55.5 gallons (210 l) fuel tank inside the dorsal fairing. In addition, to improve longitudinal stability, the E-6U second prototype received the installation of an 88 lb. (40 kg) ballast counterweight directly behind the regulating cone of the bow air inlet. However, this variant also differed from the first prototype by the installation of a sod-57 (Cross Up) transponder, a modified KAP-2 type autopilot (with automatic height and speed correction), and a PT-21 brake parachute.

The two-seater MiG-21U had a similar performance as the one-seater MiG-21F-13 in terms of stability, maneuverability, and agility, and it was approved for mass production. The State Committee, however, had some reservations regarding the MiG-21U trainer. The plane's nose tended to tilt upwards while the gear was down during landing and taking off at full speed with no afterburner. The nose-up tendency of the MiG-21U was inherited from the MiG-21F/F-13.

The State Committee also questioned the placement of a ballast in the front portion of the fuselage. Although its installation significantly boosted longitudinal stability, the Committee claimed that it also resulted in lower directional stability at certain speeds, a drop in acceleration, and increased fuel consumption.

The trainer retained the general characteristics of the MiG-21F/F-13, with the second cockpit taking up the area previously allotted in the fuselage for fuel. Consequently, the fuel capacity was reduced to 620 gallons (2,350l).

Although the MiG-21U was based on the MiG-F/F13, it borrowed several mods from the MiG-21PF. For Instance, the MiG-21U used the MiG-21PF's landing gear with enlarged wheels, smaller front airbrakes, and a double canopy hinged on the right. After receiving a few MiG-21U trainers, the Indian Air Force branded it as the 'Type 66-400' series.

The MiG-21U was also modified in other ways. The 'Mongol A' was stripped of all weapons, including the N-30 cannon and its magazine, for weight reasons. If a gun was necessary for firing practice, it was decided that an 'Afanasiev A 12.7' heavy machine gun, complete with a magazine and control system, would be installed in a pod and slung under the fuselage on the centerline pylon. Another modification was the expansion of the vertical stabilizer.

The first three serial MiG-21Us left the plant in 1962. By 1966, 180 MiG-21Us had rolled off the assembly line of plant no. 31 in Tbilisi, Georgia.

Three photos of the MiG-21U 'Red 61' (Mongol A) (courtesy airwar.ru)

This is a MiG-21U-600 of the Polish AF, c/n 662720, s/n 2720. Delivered way back in August 1966 and first served with 4PLM. In 1975 it was assigned to 1PLM until February 1989, when it was wfu. Currently preserved at Sil Powietrznych Muzeum in Deblin. (Alan Wilson)

MiG-21US (E-6US)

The MiG-21US (E-6US) trainer, 'Type 68' (NATO: Mongol B), was a new trainer version based on the MiG-21PFM that included the 'SPS' flap blowing system. However, the installation of 'SPS' was not the only modification made to the MiG-21US. Another significant change was replacing the outdated SK ejection seat with the more dependable KM-1.

The engine came next. The MiG-21U's R-11F-300 turbojet, with 12,500 lb. (6,000 kg) thrust, was replaced by the more powerful R-11F2S-300, with 13,614 lb. (6,175 kg) thrust, for the MiG-21US.

OKB MiG also modified the MiG-21US fuel system. Even though the size of the dorsal spine fuel tank had increased, its weight did not change because of a reduction in pipeline length, which resulted in fewer attachment points. This modification increased the fuel supply from 618 gallons (2,340 l) to 647 gallons (2,450 l).

Furthermore, the mechanism for supplying fuel to the combustion chamber on startup was removed. That system was obsolete and a relic of older MiG models.

Before installing the KM-1 ejection seat, the cockpit also had a significant alteration. Because the flight instructor had little opportunity to see forward from his position behind the student pilot, except for a short space between the two cockpits, OKB MiG increased the instructor's view by placing a small periscope in his canopy. The periscope rose automatically above the flight instructor's canopy when the landing gear was extended and slid back into position when the wheels retracted. This crucial modification allowed the flight instructor to see ahead during the most vital parts of the flight, which were takeoff and landing.

However, that situation did not last long. The MiG-21US and MiG-21UM trainers were equipped with the new KM-1 ejection seats. After obtaining their forward vision periscope and feeling great, the flying instructors were in for a rude awakening. The massive headrest of the KM-1 ejection seat for the student pilot entirely obscured the view of the flying instructor sitting in the back.

Moreover, the unlucky flight instructors could not fully utilize their periscopes since the headrest was too high up even for the periscope to provide a clear view. Thus, the flying instructors were forced to fly virtually blind once more, hoping for the best.

A total of 347 serial MiG-21USs were produced at plant No. 31 in Tbilisi, Georgia, between 1966 and 1970. There were two versions, the 'Type 68' for the Soviet Air Force (VVS) and the Soviet Air Defense Forces (PVO), and the export version 'Type 68A'.

This is a MiG-2US trainer of the Polish AF, c/n 01685144, s/n 4401. It was delivered in March 1970, and was assigned to CSL, at Modlin AB, then to 62 PLM. It was wfu in January 1990. It is currently preserved at the Muzeum Lotnictwa Polskiego w Krakowie (Polish Aviation Museum) in Kraków. (Alan Wilson)

This is an 'exotic' looking MiG-21US, c/n 01685141, s/n 5141. It was delivered to the Polish AF in August, 1969, and was first assigned to CSL, at Modlin AB. Later it joined 62 PLM before being wfu in May, 1991. It is now preserved at the Wanaka Transport and Toy Museum in New Zealand. (Paul Nelhams)

MiG-21UM (E-6UM)

The 'Type 69' MiG-21UM (E-6UM) was the ultimate combat-trainer variant. It was designed for third-generation MiG versions. The MiG-21SM, M, MF, and R were among them. The MiG-21UM was the final mass-production variant of the MiG-21U two-seater combat-trainer series. It was outwardly almost identical to its predecessor, the 'US' version, which is why NATO labeled it 'MONGOL B.' The most significant differences, however, were in the MiG-21UM's equipment and instrumentation.

Firstly, the new AP-155 autopilot was installed on the 'UM' instead of the KAP-2 autopilot. The MiG-21UM was the only one of the three MiG-21 combat-training aircraft to use this autopilot. Furthermore, the AP-155 was a three-channel autopilot that controlled the ailerons, and elevators, whereas the KAP-2 autopilot only controlled the ailerons.

Regarding the powerplant, early MiG-21UMs were fitted with the R-11F2S-300 13,614 lb. (6,175 kg) engine, while late production jets were equipped with the more powerful R-13-300 14,550 lb. (6,490 kg) turbojet.

The 'UM' was also supplied with modified aiming equipment by replacing the ASP-5NM type optical sight with an ASP-PFD-21 type optical sight. The ASP-FD sight was good for air targets at ranges of 1,970-6,560 ft. (600-2,000 m.) when launching unguided rockets and firing a machine gun and at 3,280-29,530 ft. (1,000-9,000 m.) when launching guided missiles.

The 'UM' was fitted with newer ejection seats. The KM-1UM (student's front seat) and a KM-1IM (instructor's back seat) replaced the KM-1U and KM-1 used in the previous version.

A DUA-3A critical angle of attack sensor was mounted on the left side of the forward fuselage, and an antenna appeared on the surface of the fairing closer to the horizontal stabilizer.

Because no training variant of the MiG-21bis was built, the MiG-21UM became the most advanced two-seater fighter in the MiG-21 series.

The MiG-21UM was serially manufactured at Tbilisi Plant no. 31 in 1971. By 1985, 1,133 of these combat trainers had been built. They were available in two versions: one for the Soviet Air Force 'VVS' and USSR Air Defense Forces 'PVO,' and one for foreign customers.

A MiG-21UM, c/n 516999351, s/n 9351, of the Polish AF. Delivered in May, 1981, its first assignment was with unit 11PLM. It was then transferred to unit 62PLM, before it was wfu from Poznan AB. In May 2004, it was preserved at 31st Air Base at Poznań-Krzesiny, where it remains to this day. (Jerry Gunner)

This MiG-21UM '21 red', was previously coded '106 red'. The reason it looks it just came off the production line is because it was refurbished by the 121st Aircraft Repair Plant at Kubinka in early 2016, according to Alan Wilson. It's now on display in Area 1 of the Patriot Park in Kubinka, Russia. (Alan Wilson)

A Romanian AF MiG-21UM, c/n 516999176, s/n 176, which was delivered in 1989. First off, it was with unit 91 BdeV. It was later converted to a Lancer-B. It joined unit 71 BdeV in 2001, and moved to EdeV 711 in 2015. There were initially 10 Lancer B trainer conversions of the Mig21UM variant in the Romanian AF. That eventually rose to 14 at the expense of two Lancer A and two Lancer C types. (Richard Vandervord)

This is a MiG-21UM of the Hungarian AF, c/n 516917011 s/n 7011. Delivered in May 1974, it served with 47 HRE, until its retirement in 2000 from Papa AB. It was preserved in 2008 at Kecskemet AB, and there it still remains. (Richard Vandervord)

The MiG Design Bureau developed an experimental fighter powered by the E-50 engine. As a consequence, the famed MiG-21 fighter was created. From that moment forward, the MiG-21 became one of the most famous aircraft in the world. Furthermore, the successful design solutions found during the aircraft's development made it possible to create numerous modifications for this very adaptable and durable fighter.

Its successes are counted in solid facts. It broke 24 world records. It was mass-produced for 30 years, and over 70 variants and modifications (major and minor) were made during its service life.

Serial production was carried out by large aircraft plants throughout the USSR, in Gorky, Moscow, and Tbilisi. Additionally, the MiG-21 was built in Czechoslovakia, India, and China. Although production ended in 1986, the MiG-21 still serves in several nations' Air Forces, including two NATO members, Romania and Croatia, (as of 2022) more than six decades after its maiden flight. On top of all that, with over 10,000 units sold globally, the MiG 21 'Fishbed' is the most widely produced jet fighter in the world. To put it in perspective, the F-16 is way back in second place with around 4,600 units sold.

Top photo: MiG-21PFM. (Aaron Headly) Bottom photo: MiG-21UM (Richard Vandervord)

THE MiG-21 IN THE VIETNAM WAR

To the reader: The Mig-21 was involved in many conflicts throughout the world. It's beyond the scope of this book to include all the wars that the MiG-21 was engaged in. Thus I decided to have the most important and most interesting war that the Mig-21 participated in; the Vietnam War.

That war was no regular war; it was a war that transcended the borders of Vietnam, not just into the neighboring countries of Laos, Cambodia, Thailand, and China, but to the United States as well.

It was a war for the soul of America, and it was a war that pulled apart the fabric of American society. But, from the large antiwar demonstrations to the political complexities and the regular soldier going to face his fate in a distant land, the Vietnam War was not a war for the future of Vietnam; it was also a battle for the future of America itself.

THE BATTLE BEGINS

The official start of the Vietnam war did not occur on land but on the sea. On August 2, 1964, South Vietnamese commandos attacked two small North Vietnamese islands in the Gulf of Tonkin. The American navy destroyer 'USS Maddox' an electronic spy ship, was 123 miles south with orders to electronically simulate an air attack to draw the North Vietnamese (NVN) boats away from the South Vietnamese commandos and give them a chance to escape.

Suddenly, three fast NVN torpedo boats, cutting through the sea at top speed, attacked the Maddox scoring a hit with a single machinegun bullet but missing with the torpedoes. Gunfire from the destroyer, and attacks by F-8E 'Crusaders' from the aircraft carrier USS Ticonderoga, sank one of the torpedo boats and badly damaged another. Following this action, the USS Maddox was joined by another destroyer, the USS Turner Joy. Both ships resumed patrol to gather information and underscore the United States' insistence on the right of free passage in international waters.

Then, on the night of August 4, while the two destroyers were on patrol, three NVN torpedo boats appeared, shadowing the American warships before closing in at tremendous speed. Two assault boats were reportedly sunk, and one was severely damaged in a fierce fight that lasted well past midnight. The two destroyers were not damaged.

As a result of the second attack, the US Congress enacted the 'Tonkin Gulf Resolution,' allowing the President of the United States the right to take whatever steps he thought necessary to discourage further communist aggression in Southeast Asia. President Johnson then directed the deployment of US Air Force aircraft to Southeast Asia in the event of North Vietnamese or Chinese reprisal for the carrier strikes.

Following the Tonkin incident, three further attacks on American and Vietnamese military posts in SVN in late 1964 and early 1965 prompted the retaliatory operation 'Flaming Dart' air attacks against NVN.

The NVN possessed just 25 Chinese-made J-5 subsonic fighters, numerous Soviet MiG-15bis, and Il-28 two-engine bombers, similar to the North American B-45 Tornado, at the start of the Vietnam War. These three models could not compete with the Americans' superior jet fighters. Indeed, the NVN AF was hesitant to use its jet fighter force for anything other than defensive patrols. As a result, even though no communist assault aircraft attempted to target South Vietnam, USAF air defense units made forays into North Vietnam, where they engaged in air-to-air combat with the NVN AF.

In the early months of 1966, before the introduction of the MiG 21, 11 MIGs were shot down, resulting in the loss of only three USAF aircraft. The MiGs were primarily used for battle training, with American Ground-Controlled Intercept (GCI) controllers positioning the interceptors for direct attacks. The MiGs completed a dry firing pass and usually broke off before engaging in combat with the USAF fighters. However, this operational training aided in integrating the GCI/MIG system into a mature interceptor capability.

For the USAF, the main danger during the early stages of the conflict was not the MiGs, but active antiaircraft artillery and automatic weapon (AAA/AW) defenses, which were eventually integrated with the Surface to Air Missile (SAM) and MiG defenses. The MiGs got lucky when on April 4, 1965, the first two F-105 Thuds were shot down by NVN AF MiG fighters. That was before the USAF confirmed the destruction of its first MiG on July 10. On the other hand, the USN beat the USAF by downing its first MiG on June 17 that year.

March 1965 saw Washington inaugurating its highly restricted and carefully controlled series of numbered 'Rolling Thunder' air campaigns against the north.

Still, in the early days of the conflict, the USAF could have done far better. By early 1966, there was no doubt about the USAF's air supremacy; nonetheless, considerable enemy counter-air capability remained. Moreover, the Rules of Engagement (ROE), which were politically driven, had a significant (negative) role in the war. For example, US fighters and bombers were not authorized to strike airfields, render them unusable, or destroy aircraft on the ground. Try telling that to the Israeli Air Force before the Six-Day War in 1967. In addition, the ROE prohibited striking enemy fighters unless they were endangering US forces first. With such political constraints, the USAF was essentially fighting with one hand tied behind its back. The limitations these regulations imposed on commanders and individual fighting men proved to be a frustrating and costly example of micromanagement at its most extreme.

Then in April 1966. the MIG 15s and 17s were strengthened with the arrival of the MIG 21s. Some were upgraded and equipped with infrared homing missiles, the K-13, a copy of the US Sidewinder missile (NATO CODE: 'Atoll'). Suddenly, the American Forces' control of the air over Vietnam appeared to be under threat. The Phantom F-4's major adversaries were now the supersonic MiG-21F-13 (some of which were Czechoslovakian-made) and the MiG-21PF all-weather interceptor, which was outfitted with the dependable RP-9-21 (CD-30T) (Spin Scan) radar sight. They also sported a 'SEA' tropical camouflage, similar to the American aircraft, and had TGS (rocket pods) with 55-mm S-5 unguided aircraft rockets.

The US Air Force and the US Navy had high aspirations for the F-4 Phantom, believing that

The strong and robust F-4 Phantom, would be a thorn in the side of the MiG-21, and vice versa. (USAF)

powerful weapons, enhanced airborne radar, high speed, and acceleration characteristics, combined with novel tactics, would provide the Phantoms with superiority over enemy aircraft. However, upon engaging with the lighter MiG-21s, the F-4s began to suffer losses. Disputably, the NVN claimed that from May to December 1966, the US side lost 47 aircraft in air battles while destroying only 12 enemy fighters.

The F-4's extensive wing loading, somewhat lower angular turn speeds, particularly at medium altitudes, poor handling, and restricted angle of attack limited its operational overload to 6.0 v 8.0 for the MiG-21PF. The F-4 also had no advantage in terms of thrust-to-weight ratio. The F-4B had a typical takeoff weight of 0.74, while the MiG-21PF weighed 0.79. Another drawback of the 'Phantom' was its poor spin characteristics. The

A North Vietnamese MiG-21PFM preserved at the Barksdale Global Power Museum. (Michael Barera)

jet might go into a flat tailspin, which a pilot with ordinary ability might not be able to recover from. Before 1971, 79 Phantoms were claimed to have been lost in the Vietnam War due to tailspin. Apart from aerodynamics, the F-4's radar possessed excellent detection, capture range, and low noise immunity. Numerous instruments and controls surrounded the pilot and operator's cockpits. In that regard, Russian fighters were years behind.

The pilots of the NVN AF's oldest unit, the 921st Fighter Aviation Regiment (IAP) named 'Sao Do,' earned the distinction of being the first to master the MiG-21 jets. Some pilots, including future aces Nguyen Van Kok, Pham Thanh Ngan, and Nguyen Hong Ny, were trained in the Krasnodar Higher Military Aviation School of Pilots in the Soviet Union.

Before making the first flight on the MiG-21, any cadet in the Soviet Union had to master the first combat jet trainer, the Czechoslovak-built L-29 Dolphin. If successful, they would be trained on the supersonic MiG-21U combat trainer. Unfortunately, the NVN lacked such aircraft, so initial training was conducted on the subsonic MiG-15UTI and sometimes on the JJ-5 (the two-seat variant of the Chinese-made MiG-17 copy).

Back in North Vietnam, the retraining on the MiG-21 was completed at the end of January 1966. Thus the 921st IAP was finally equipped with the newest MiG-21F-13 and the outdated MiG-17. The pilots were almost ready for battle.

The NVN MiG-21 pilots began their combat work hunting down American unmanned reconnaissance target drones, the Ryan 'Firebees.' The first such aircraft was shot down on March 4, 1966. At 4 am, the radar detected an aerial target approaching from the northeast. The deputy commander of the 921st IAP, Tran Han, gave Nguyen Hong Nyi the order to take off to intercept. After 19 minutes, the MiG-21 gained an altitude of 52,490 ft. (16,000 m); the target at that moment was 9.3 miles (15 km) away from the Vietnamese fighter and 6,560 ft. (2,000 m) above it. The unmanned reconnaissance drone was heading for Quang Ninh Province. The ground command post confidently controlled the pilot's actions, and a few minutes later, the MiG-21 fighter opened its combat account in the skies of Vietnam. The following day, another unmanned recon drone was shot down by a MiG-21 fighter from the 921st IAP. These were generally the first American unmanned reconnaissance drones shot down by Vietnamese fighters. The next step was not going to be so easy.

In April 1966, MiG-21 planes engaged American jets numerous times, but the MiG pilots' were out of luck. Then, on April 23, a pair of MiG-21s came across a flight of 'Phantoms' and started an attack, sparking the first air war. Despite the surprise advantage, the Vietnamese could not launch their K-13 infrared missiles. At least 14 missiles were launched from the MiG-21s in April and May; however, none hit their targets.

Moreover, the 921st regiment suffered the first aircraft losses when many MiG pilots were forced to eject owing to a lack of fuel. On April 26, the USAF's 480th Squadron of the 35th Tactical Air Wing claimed to have shot down one MiG-21.

The impact of the USAF's use of Electronic Counter Measures (ECM) contributed to diminishing the NVN's defensive capabilities. However, thanks to the MiG-21s, the NVN AF's aggressiveness increased in the second half of 1966. Except for four days, the MIGs flew almost every day from September 4, 1966, to January 1967, marking the first continuous use of these aircraft as active air defense weapons.

This is a MiG-21PFL, c/n 771914, s/n 4128. It was built in 1965, and served with the illustrious 921st 'Sao Do' Regiment. (mark6mauno)

When the NVN AF examined the April-May air battles in which the MiG-21 participated, it turned out that the pilots had difficulty working with the fighter's sighting system. When looking for a target, the pilot used an onboard radar; once the target was detected, he had to switch to an optical sight, from which he could calculate the rocket launch time. The pilot of the enemy aircraft attempted to avoid the strike in every manner imaginable. In order to keep the maneuvering target in the narrow field of view of the optical sight, remarkable flying skills were required, which the Vietnamese pilots did not have in the spring of 1966.

An analysis of the first battles showed that the most effective method to get a kill would be the use of pairs of MiG-21 fighters with mixed weapons. The lead fighter would be armed with two K-13 missiles with infrared homing heads and the wingman's aircraft with a couple of NAR units (rocket pods) with S-5M rockets.

This method seemed to have worked. A pair of MiG-21 fighters with mixed weapons got the first NVN AF MiG-21 kill in a dogfight. On June 7, 1966, pilots of two MiG-21PFs spotted two F-105 Thunderchief fighter bombers flying at an altitude of only 1,640 ft. (500 m). The wingman launched unguided missiles from a distance of 4,920 ft. (1,500 m), but the pilot of the Thunderchief noticed the launch in time and dodged the NAR rockets with a left turn. However, the MiG was right behind the 'Thud'; two more volleys followed from distances of 1,640 ft. (500 m) and 655 ft. (200 m). According to the Vietnamese version, the missiles found their target, but the USAF did not confirm the loss. Meanwhile, due to the significant maneuvering of the second Thunderchief, the pilot of the leading MiG could not take a position to launch his missiles.

Two more similarly scripted aerial battles occurred over the next few days. Again, the Vietnamese might well have been more fortunate if their MiGs had cannons fitted. Unfortunately for the MiG pilots, at an early stage of the war, Vietnam received 'clean' MiG-21PF interceptors that were not equipped with GP-9 cannons.

Although MiG pilots looked to be inexperienced in aerial combat, the slow intensification of the air war provided them time to shape their force into a more serious threat, as seen during the summer and fall of 1966. The MIG-17s and MIG-21s had contrasting roles, with the former focusing on low-level interceptions and the latter operating at high altitudes. Moreover, some MiGs still tried to interdict US strike aircraft during their bombing runs; others assumed positions to threaten American planes on their way to the targets. The MiGs aimed to force strike planes to jettison their munitions before they got there. The MiG pilots also discovered they could successfully outmaneuver most US air-to-air missiles with a rapid turning descent. Since the Sparrow (AIM-7) and the Sidewinder (AIM-9) had been designed to down bombers, the missiles could not maneuver fast enough in a fighter-to-fighter engagement.

As a result, US pilots requested that cannons be mounted on their F-4s. Subsequently, external 20-mm gun pods were placed on F-4Cs and used in battle for the first time in May 1967. However, until the modification was completed, MiG attacks on US strike forces proved extremely difficult to control. Only 107 sorties (2.72%) of the 3,938 strike sorties flown from September to December 1966 were intercepted by MiGs. However, 107 (55.73%) of the 192 strike aircraft engaged by MiGs had to jettison their ordnance. That clearly indicated that the MiGs minimized the impact of US strikes on the days when they were carried out.

On July 7, 1966, Vietnamese radars detected a group of American planes approaching from Thailand. A pair of MiG-21s, pilots Nguyen Nyat Chieu and Tran Ngoc Xiu, took off from Noibai AB, whose pilots were ordered not to engage in combat. However, the attack order was issued after a group of F-105D Thunderchiefs crossed the Tamdao ridge. Tran Ngoc Xiu launched two missiles simultaneously, striking an F-105D. (the USAF did not confirm the loss of the Thunderchief). Then, on July 11, another F-105D Thunderchief (c/n 61-0121) piloted by Major W.L. McClelland of the 355th Tactical Fighter Wing was shot down by a pair of MiG-21s flown by pilots Wa Ngoc Dinh and Dong Wan Song. However, the USAF again disputed the loss of the F-105D in an air battle. Instead, it reported that the 'Thud' allegedly crashed after using up all its fuel. On July 14, the USAF exacted vengeance for the downed F-105D. On that day, two MiG-21s were shot down by an F-4C fighter from the US Air Force's 480th Squadron of the 35th Tactical Fighter Wing.

In the second half of 1966, several air battles took place with the participation of the MiG-21 fighters. First, on September 21, a MiG patrol attacked a group of 'Thuds' flying under the cover of Phantoms. The leader of the Vietnamese pair shot down one F-105D with a K-13 missile from a distance of 4,920 ft. (1,500 m). The USAF attributed the loss to flak fire. Then on October 5, the USAF credited a MiG-21 for the loss of a Phantom F-4B (s/n 64-0702) from the 433rd squadron of the 8th tactical fighter aircraft wing. However, the NVN AF did not confirm this loss. Instead, it said that the MiG-21 did not shoot down the Phantom and was most likely shot down by anti-aircraft fire.

As luck would have it, three days later, on October 9, NVN MiG-21 pilots scored their first victories over US naval aviation pilots. They claimed to have shot down two Phantoms. One of the Phantoms was an F-4B (BuNo 152093) from USN fighter squadron VF-154 'Black Knights' of Carrier Air Wing 2 aboard USS Coral Sea (CVA 43).

The crew of the Phantom was pilot Lt. commander Charles Tanner and avionics operator Lieutenant Ross R. Terry. They managed to eject safely, but unluckily, once they landed, they were captured by the NVA. The two men spent six and a half years in captivity. They were subjected to physical and psychological torture in order to disclose military information. The NVN also forced them to admit that a Vietnamese fighter shot them down. According to the US Navy, the Phantom with Charles Tanner and Ross R. Terry was the only naval aviation jet lost on October 9, 1966. The fact that the NVN forced the pilots to admit they were shot down by an NVN jet was because it probably wasn't. It was most likely shot down by anti-aircraft fire, not by an air-to-air missile.

On the same day of the loss of the F-4B Phantom, a USN pilot shot down the first MiG-21. The downing of the '21' was achieved by squadron Commander Dick Bellinger of Fighter Squadron 162 'Hunters' aboard USS Oriskany CVA-34 flying an F-8E Crusader. Then on November 5, Phantoms from the 480th squadron of the 366th

USAF Tactical Fighter Wing reported two more downed MiG-21s.

Several air battles occurred in the first half of December 1966. The first came on December 2,

This October 9, 1966 photo, shows Commander Richard Bellinger, next to his F-8E Crusader, talking about shooting down the MiG-21. He was the first Navy fighter pilot to down one of the new MiG fighters over Vietnam. (Official U.S. Navy Photo 'USN 1121723')

when a group of Phantoms and 'Thuds' focused their attention on the Noi Bai AB. Three MiGs took off to defend the base. The MiG pilot of the first pair successfully launched a K-13 missile at one of the Phantoms from a distance of around 4,500 feet (1,370 m) and scored a hit. His wingman, however, was unable to capitalize on his commander's success because the distance required for a successful strike with his unguided rockets was too great. Another Phantom, F-4C, was shot down by the leader of the second pair of Mig-21s. When the third pair reached the area, the American fighters were already out of range of the K-13 missiles. The USAF confirmed the loss of the two F-4C Phantoms on December 2, but the losses were attributed to SAMs.

Until the end of 1966, Vietnamese fighters operated only in the area around Noi Bai AB. The MiG-21s only engaged hostile aircraft if they attempted to breach the airspace surrounding Noi Bai. Russian advisers with the NVN urged that the Vietnamese modify their tactics. They suggested attacking the Americans when flying in a close formation on their route to their objective rather than waiting for them to violate the base's airspace. The NVN took notes, and the new tactic was tested for the first time on December 5, 1966.

Two MiG-21 planes took off from the Noida AB and retreated to a distance of 22 miles (36 km) from the Air Base. A group of 24 Phantoms and Thunderchiefs reached the location after a few minutes. The first Vietnamese fighters engaged and shot down the leading F-105D 'Thud.' The USAF confirmed the loss of the F-105D (s/n 62-4331) from the 388th Fighter Wing's 421st Tactical Fighter Squadron. However, it claimed the 'Thud' was shot down by MiG-17 gunfire, not by a MiG-21. The pilot, Major B.N. Bigley, was killed.

THE CONTINUING THREAT - 1967

In December 1966, Mig-21s shot down nine USAF aircraft (while losing 25 of their own) and, by the rules of the ROE, continued to have relative immunity from attack except when they chose to engage in battle. The NVN AF Order of Battle (AOB) for December 15, 1966, showed 53 MiG 15s, 17s, and 21s on station at Air Bases Phuc Yen and Gia Lam. It noted that Communist China had nearly 500 MiGs at its disposal, which could readily be transferred to the NVN AF or used in NVN AF/ChiCom (Chinese Communist) AF operations.

The New Year began badly for the NVN AF. The US Air Force opened the Year with 'Operation Bolo.' The operation was devised by Col. Robin Olds, commander of the 8th Tactical Fighter Wing (TFW) and the wing's tactics officer, Capt. John 'J.B.' Stone. Their brilliant strategy was to attract and trap North Vietnamese MiG-21s by mimicking an F-105 bombing formation with their F-4 Phantoms to ensure that the North Vietnamese would engage in an air battle, a force of Forward Air Controllers (FACs) would simulate an impending F-105 and an F-4C strike. During the Vietnam War, FACs were rated pilots who coordinated air-ground operations. FACs may be

The legendary Col. Robin Olds on the left, with his trademark handlebar mustache, and one of four NVN stars on his Phantom, denoting four kills. On the right he is celebrating with his men after completing his 100th combat mission. (USAF)

helping ground forces or flying over the conflict zone. It should be noted that their aircraft were not designed for combat but rather to aid air operations by ensuring that the fighter and fighter bomber strikes hit their intended targets while avoiding injuring friendly forces.

When their radar picked up the 14 'F-105Ds', the NVN Air Force must have felt they hit the jackpot. As expected, a large MIG-21 force, presumably manned by newly trained Vietnamese pilots fresh from training at the USSR, took off to confront what they thought were predominantly F-105 fighters and their crews. Consequently, seven MIG-21s were destroyed within 12 minutes of combat, with no USAF aircraft losses. Then, on January 6, F-4C Phantoms simulated a weather reconnaissance mission, enticing the MiG-21s into battle. This time the consequence was the destruction of two additional MIG-21s. One NVN pilot ejected, and the second one died. After being stunned by their losses, the North Vietnamese Air Force paused for further training until February 1967.

On January 8, 1967, the NVN Air Force Command convened to discuss the 921st IAP's heavy losses. It was decided to switch tactics once more. The MiG-21 would need to engage the enemy in a hit-and-run guerrilla warfare style. For each mission, two to four fighters were to be deployed. When approaching the enemy, the pilots were advised to use confusing maneuvers such as constantly changing altitude and speed. An attack from behind or above was thought to be preferable. For example, when the NVN AF MiG-17s noticed the F-4s turning to attack the MiG-21s, they would attack the Phantoms from below and the flanks.

With the northeast monsoon hampering operations from January to March 1967, the number of attack sorties declined. Then, things started to get intense in the spring. The 'harassment and disruption' of Hoa Lac and Kep Air Bases, as well as second strikes on Kien An in April, May, and June 1967, resulted in the destruction of 40 MiGs in the air and on the ground.

However, US losses during the same period totaled an astounding 98 aircraft. It was the highest quarterly toll in the air war, including nine more aircraft lost in May to 431 SAM attacks (usually fired in a barrage), which was the greatest number of SAM missiles to be fired in a single month. By June 7, all operational MiGs still in the NVN AF were concentrated on Phuc Yen and Gia Lam, while the others sought sanctuary in China.

Nonetheless, the MiG problem was causing all types of difficulties for the American side. On May 2, the Commander-in-Chief Pacific Command (CINCPAC) wired the Joint Chiefs of Staff (JCS) to convey his concerns about the 'MiG menace.' He stated that the number of air engagements during recent strikes against JCS-assigned targets was indicative of the increasing MiG threat to the US forces. He thought attacks against the remaining jet-capable airfields of Phuc Yen, Kien An, and Cat Bi should be considered necessary to harass further and disrupt the MIG air defense capabilities. He concluded by saying that the primary goal of the strikes would be the 'destruction' of MiG aircraft. A secondary objective would be to disrupt the support facilities associated with the bases.

A few days later, the Pacific Fleet Commander-in-Chief (CINCPACFLT) warned that the NVN would go all out to defend its threatened MiG fleet. Instead, CINCPACFLT suggested diversionary strikes to flush out the MiGs, followed by another strike 30 to 40 minutes later to inflict significant damage on the MiGs.

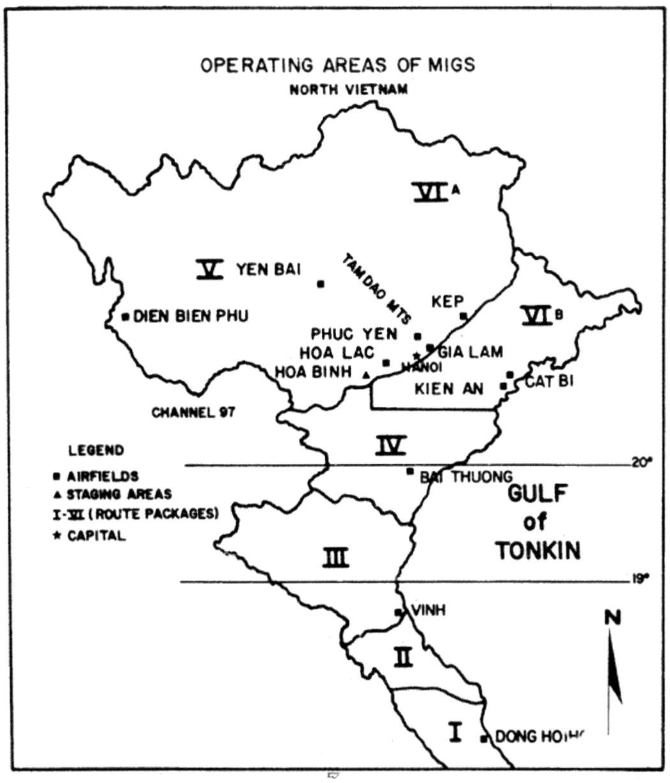

NVN ROUTE PACKAGES AND KEY AIRFIELDS

(Map: USAF)

The Pacific Air Forces Commander-in-Chief (CINCPACAF) fully agreed and ordered maximum strikes, suggesting that attacks be spaced 20 - 30 minutes apart to catch the MiGs on the ground. Subsequently, the strikes were made, with 26 MiGs destroyed in May alone. But, there were severe consequences; the US forces lost 40 aircraft in all Route Packages.

Also in May, American planes resumed their raids on Hanoi. During this time, fighters from the NVN Air Force began flying up to 40 sorties daily. Colonel Robin Olds shot down a MiG-21 on May 4. Eight days later, the 921 IAP pilots exacted revenge when, over Vanyen, a pair of MiGs flown by Ls. Trong Huen and Dong van Song shot down a two-seat Thunderchief, which the USAF later attributed to anti-aircraft artillery, not the MiGs.

For the first six months of 1967, the Americans became more proficient in the air battles, as was seen in 'Operation Bolo.' In May, the US forces had several air battles with the MiGs in the Hanoi-Haiphongar-area, situated in the North East of North Vietnam, designated by the US forces as 'Route Package VI' (see map on the left). However, the NVN AF planes were not all Mig-21s; there were plenty of MiG-17s as well. The May total of all 72 engagements was 26 NVN losses against 2 US losses, which reflected the intensity of the USAF's effort and its superiority in air-to-air combat. After losing five more MIG-17s early in June, the NVN AF again stood down to take a fresh look at the situation. In less than six months, they had lost 51 MiGs in the air and some 30 on the ground. The losses were equal to the most optimistic estimate of their inventory of MiGs in both NVN and Communist China.

In terms of losses, US pilots had other things to worry about since the start of Operation 'Rolling Thunder' than MiG attacks. For example, during a period of reduced MiG activity, July-September 1967, the US lost three planes to MiGs, but 16 to SAMs and 73 to anti-aircraft artillery/automatic weapons (AAA/AW). When one looks at the situation, AAA/AW was the biggest threat, followed by SAMs and, lastly, the MiGs. The attrition rate for the US aircraft from SAMs and AAA/AW attacks was utterly unacceptable. Yet, the chase for MiG kills continued. A total of 401 aircraft were lost in 1967 alone, the second worst year after 1968. Only 16 were lost to MiGs out of that number. (For a table of US aircraft losses, see the following page). Incidentally, in 1969 and 1970, no US aircraft were shot down by SAMs.

By late 1967, US strategies against the MiGs had finally become relatively fixed, while the NVN AF was always trying to develop and perfect techniques to counter the ' Rolling Thunder strikes.

The USAF had Specific instructions from the tactical doctrine for fighting the MiG-21s. The two pilots were to make the attack in a 'Fluid Four' formation. This formation tended to box the MiGs in and usually placed one of the jets close to the MiG pilot's rear side blind area. Both jets had to work to force the fight to an altitude below 15,000 ft. The two pilots would need to continue to work as a flight and constantly push the fight as low as possible to maneuver toward

YEAR	GROUND FIRE AAA/AW	OPERATIONAL	SAM MISSILES	GROUND ATTACKS ON US BASES	AERIAL COMBAT	OTHER COMBAT	TOTAL
US AIRCRAFT LOSSES DURING THE VIETNAM WAR							
1962-66	298	125	9	21	4	—	57
1967	276	77	26	6	16	—	401
1968	275	91	19	45	22	10	462
1969	234	77	—	14	—	1	326
1970	177	76	—	3	1	—	257
1971	73	27	2	1	—	—	103
1972	72	19	24	2	14	1	132
1973	38	25	30	4	10	9	116
1974	—	1	—	—	—	—	1
TOTAL	1,443	518	110	96	67	21	2,255

(Source: USAF Operational Summary)

the MiG pilot's blind area. The pilots were warned not to attempt a horizontal turning fight but to use the vertical to gain a firing position. If the MIG-21 had the advantage, the pilot was to disengage immediately. He will need to make a break into the attack if a missile has been launched. A hard turn would be enough if the MiG was only approaching the launch envelope. Then continue with a slicing descending turn to drag the dog fight to a lower altitude. If the MiG attempted to continue the attack, the supporting jet should be able to 'sandwich' him. If no other jet was available, the lone pilot was to accelerate to max speed at low altitude and disengage.

With these techniques, US fighter pilots felt comfortable mixing it with the MiGs whenever they might reappear in force. The MiGs didn't take long to show up. As the Northeast monsoon arrived, the number of US attack sorties per month steadily decreased from 11,744 in August to 5,758 in December. In contrast, the number of MiG engagements increased from 16 in August and September to 34 in December. Thus, strike sorties were cut in half, while engagements doubled.

The NVN AF lost a total of 20 MiGs during the crucial month of October. Two MIG-21s, six MIG-17s, and twelve more were shot out of the sky by US pilots. On October 24, MiGs were destroyed or damaged during a newly authorized strike on Phuc Yen Airfield. The overwhelming air superiority that American pilots had fought so hard to achieve earlier in the year had to be maintained. Only three US fighters were shot down by the MiGs.

Based on previous experience, the NVN AF was expected to pull back, but this time they persisted and managed to gain a slight advantage in the air battles over the following two months. In air-to-air combat, the total score was the loss of six MiG-17s to nine US losses; four F-105s, two F-4Bs, and three F-4Ds.

Despite their drastic October losses, the NVN AF managed to secure enough replacements that their MiG inventory rose from 75 to 100 by the end of the year.

THE YEAR OF TET- 1968

The year 1968 began somewhat peacefully, but by the end of January, the 'Tet Offensive' began. The NVN and Viet Cong worked together and launched a massive coordinated offensive against US and allied forces across SVN. The Viet Cong, also known as the 'Viet Nam Cong San' (Vietnamese Communists), was a guerrilla force that participated in insurgent operations within SVN. During the offensive, fierce fighting broke out across the country as the NVN and the Viet Cong attacked many cities and towns throughout SVN. As a result, American forces frequently referred to them as 'VC.'

The 'Tet Offensive' marked a turning point in the Vietnam War. Although the offensive was a costly defeat for the NVN and VC forces, the offensive was a stunning propaganda victory for the communists. In fact, it is often credited with turning the war in their favor by causing the South Vietnamese population and US citizens at home to question the war's outcome. 'Tet' also contributed to the formation of a significant peace movement in the United States. Although the 'Tet Offensive' had little or no effect on the air war, it altered the course of the war from that point forward. The map below depicts the various locations attacked simultaneously during the 'Tet Offensive.'

(Map: USAF)

THE LAST OF THE BIG AIR-TO-AIR WAR

During the first quarter of 1968, there was a significant decrease in air activity over NVN. Mainly because of the New Year and Tet stand-downs, but mostly because the Northeast Monsoon was in full swing. Compared to the last three months of 1967, the number of combat sorties was cut in half. As a result, SAM firings and kills, MiG engagements, aircraft losses, and AAA/AW activity decreased. MiG pilots appeared more confident following their recent victories, but US pilots were determined to maintain air superiority.

February had the worst flying conditions in the preceding three years due to monsoon weather, which was not a good time for dogfights. As a result, the US resorted to bombing. Most aircraft had all-weather bombing equipment to undertake medium and high-altitude strikes. For example,

1) the 'COMMANDO CLUB' missions (radar-guided bombing missions employed in SEA that used the MSQ-77 radar system to guide the aircraft to the bombing target).

2) 'COMBAT SKYSPOT' missions (ground-directed bombing system, also utilized in 'BARREL ROLL') (an interdiction and close air support operation, under 2AD control, which began on December 14, 1964, in Eastern Laos, and was reduced on April 3, 1965, to an area of northern Laos, and portions of NVN). For 'Combat Skyspot, 'Ground controllers used radar and bombing computers to direct aircraft to the bomb release sites.

3) Regarding the 'COMMANDO NAIL' missions, they performed night and all-weather bombing operations, which were carried out by aircraft outfitted with an airborne radar bombing system. The technology was similar to the US Navy's 'A-6 blind bombing' technique. As a result, MiG encounters were few and far between, with just over one every two days. On the other hand, the MiG inventory increased to 18 in

NVN and 100 in Communist China, but they were used sparingly.

However, when used, the MiGs were effective, downing 10 US aircraft while losing nine of their own. All kills occurred during the first two months of 1968. There were only five engagements in March, with no losses on either side. The MiGs' boldness was inspired by their latest victories. MiG-21 pairs performed repeated runs on US MIGCAP and attack aircraft, and solo planes thundered down on the US fighters, which didn't have the benefit of ground control.

This first narrative, with a twist, gives an illustration of the MiG pilots' aggression. On January 3, a swarm of 48 American planes appeared in the skies over Mai Chau province, moving towards Thanh Son. Two MiG-21s, with pilots Nguyen Dang Kinh and Bai Dak Nyo, took off from the Kep airstrip to intercept. The NVN pilots attacked from the sun's direction and shot down one F-105 Thunderchief on the first run. The euphoria of triumph, however, was considerably dampened by an accident on the landing of Nguyen Dang Kin's aircraft. In his haste to celebrate his kill, the pilot made a hard landing, breaking the MiG's nose landing gear in the process.

In the same battle, two other pilots, BaiDak Nyo and Ha Wan Chak attempted to intercept the 48 US aircraft but could not reach their target. Then, at about 3 pm, a group of 36 American planes was seen flying towards Hanoi. At 3.16 pm, a MiG-21 fighter piloted by Ha Van Chak took off to intercept the 'Thuds.' The pilot rapidly reached the Thud's height of 18,000 feet (5,500 m). Before Chak was in position to attack, four Phantoms swooped down on him, so he continued to climb. The Phantoms fell behind as Chak ascended to 32,800 feet (10,000 m). Then, with the Thunderchiefs flying far below him, he flew down at such a breakneck speed toward them that he barely had time to aim as a Thud whistled past him. Chak figured out another way to reach the Thunderchiefs and attempted to attack them, but again, he couldn't get into a position to launch a rocket. Meanwhile, the MiG had 185 gallons (700 liters) of fuel left in its tanks, which was less than the required amount, so he flew back toward his home base. Suddenly, out of the blue, he noticed that the direction he was flying was the same as the eight Thuds. With little to lose, Ha Wan Chak launched the Soviet sidewinder (K-13 'Atoll' missile) from 2.2 miles away (3.5 km). The Vietnamese reported the downed Thunderchief, but the Americans did not confirm the lost Thud. The fact that Chak felt glad that he had stopped an attack on Hanoi with his exceptional maneuvering skills was more relevant than whether the Vietnamese pilot had shot down a USAF fighter bomber.

There were more successes for the NVN pilots. Nguyen Dang Kinh and Dong Van Song of the 921st Regiment shot down a second EB-66 after the NVN claimed to have destroyed the first EB-66 on November 19, 1967. The United States Air Force did not confirm that loss. This time, however, the USAF verified the loss of an EB-66C s/n 55-388 from the 355th Tactical Fighter Wing's 41st Electronic Warfare Squadron, which was shot down by an infrared K-13 'Atoll' missile on January 14, 1968, near Ban Xan, North Vietnam. (The aircraft was originally a B-66B-DL before being converted to an EB-66C). On March 3, NVN pilots Nguyen Dang Kinh and Nguyen Dac Thuan claimed a third EB-66 kill. On that particular day, however, US forces again denied losing an EB-66.

Another encounter in February favored the US jets. A flight of four F-4 MIGCAP Phantoms from Ubon, Thailand, emerged out of the clouds when a MIG-21 suddenly appeared on the scene, making a high pass on the Phantoms from the rear quarter. The flight split off and pursued the MiG. Three F-4s missed it with missiles; however, the fourth F-4 found the MiG directly in front of him. The pilot immediately acquired full systems lock-on with interlocks and guided an AIM-7 semi-active radar-homing air-to-air missile straight into the MiG.

Despite its primary role as a bomber, the F-105 had considerable success against the MiGs. Its pilots used the aircraft's internal cannon to shoot down just under 30 MiGs. The firing port can be seen in the photo's upper left corner. (USAF Museum)

MiG-21PFL, c/n 771915, s/n 4228, of the NVN Air Force, with its drogue parachute. This aircraft was delivered to the North Vietnam AF in 1966 and served with the renowned 921st 'Sao Dao' Regiment. (USAF Museum)

This was an RB-66B-DL before being converted to an EB-66E. At least seven USAF Douglas EB-66E ECM aircraft were based at Takhli Royal Thai AFB during the Vietnam War. This EB-66E, seen here in December 1968, was assigned to the 41st Tactical Electronic Warfare Squadron, 355th Tactical Fighter Wing. (USAF)

ENDING THE BOMBING

On March 31, 1968, President Lyndon Johnson announced that 'Operation Rolling Thunder' (the bombing of North Vietnam) would end at 0800 hours on April 1, 1968, except for the DMZ.

Moreover, President Johnson advocated beginning peace negotiations. Aside from the aim to try to stop the war by encouraging peace, the decision had a political angle. Knowing he would not run for President in the November 1968 elections, Johnson was concerned that Richard Nixon, the Republican Party's strongest presidential rival at the time, would be the nominee. As a result, Johnson needed to assess whether the bombing halt decision would favor Hubert Humphrey, the most impressive Democratic challenger for the presidency, or disadvantage him. In the end, he obviously believed that announcing the end of the NVN bombing campaign would assist Humphrey's candidacy. However, President Johnson was way off the mark. Richard Nixon won the election and became the 37th President of the United States.

Back in the Vietnam minefield, the bombing limits drastically altered the nature of the air-to-air conflict. After April 3, 1968, MiGs flew south of the 19th parallel under radio and radar silence. They kept up their high-speed, hit-and-run tactics, but they usually fled north of the 19th parallel after making single firing passes.

The US Forces in SVN were informed of two restrictions after the bombing halt. The initial restriction, enacted on April 1, 1968, prohibited all bombing 'North of 20 N latitude', which was around the middle of Route Package IV. The second limitation, announced two days later, shifted the line 'South to 19° N latitude' and limited strikes to 'Route Packages' I, II, and the southern one-third of Route Package III. Thus, practically all of NVN became a haven for MiGs. Vinh and Dong Hoi were the only jet-capable airfields south of the front line, but the NVN Air Force had never used them for MiGs. Moreover, the NVN AF began moving his MiG fighters back to North Vietnam from Communist China.

The bombing limits drastically altered the air-to-air conflict. However, the US sortie rate rose steadily from an average of 5,000 per month in the first three months of 1968 to 11,931 in October. There were only 18 encounters between MiGs and US fighters during the remaining nine months of the year.

Undoubtedly expecting a return of US attacks in Route Packages V and VI, the NVN Air Force spent all of April and most of May repairing their Northern airfields. Moreover, It was planning major attacks in SVN. D-Day was set for May 23.

The NVN AF would experience what the USAF previously experienced in its attacks on NVN. Thus, when the North Vietnam AF swept south on May 23, one MiG-21 was instantly destroyed by a US Navy 'Talos' ground-to-air missile. In the meantime, below '19° N,' the MiGs encountered an array of US fighters, US Navy SAMs, minimal GCI support, and VHF radio channel jamming. During the 18 engagements, US pilots shot down six MiGs, while only two F-4 Phantoms were lost.

After the May 23 debacle, the NVN Air Force remained on its side of the fence and exploited its sanctuary 'North of 19°' to send the MiGs to perform sporadic raids south of the line in June. The NVN AF lost one MiG-21 and downed two F-4s in aerial combat. However, they lost a MiG-17 and a MiG-21 on July 8 and 9. In typical fashion, the NVN AF stood down to evaluate its tactics once more.

These are some of the skirmishes that occurred in 1968 after the April 1 bombing halt. On July 29, the NVN AF resumed its sweeps but remained cautiously above '18° 30' N' this time and lost one MiG-17. The NVN Air Force returned for another SVN attack on August 1 and lost a MIG-21. On August 25, an exchange of fire resulted in no damage to either side. Then, on September 17, two F-8 Crusaders of the United States Navy engaged two MiG-21s near Vinh. Again, there were no battle losses, although one F-8 ran out of fuel before returning to its carrier and crashed.

Two days later, two MiG-21s intercepted four F-8s pursuing a flight of MiG-21s, but one of the MiG-21s was shot down by the F-8s. Finally, on September 22, during an air battle between F-8s and MiG-21s near Vinh, one MiG was shot down by a US Navy Talos SAM, and the remaining MiGs left the area. In October, radar sometimes detected a MiG below '19° N', but American pilots could not go closer than 52 miles. Upon the cessation of all bombing of NVN on November 1, by the US President's decree, the NVN Air Force transferred GCI and air surveillance equipment to Vinh as a precaution. By December, their MiGs had been reduced to pursuing US 'Bumpy Action' reconnaissance drones near Hanoi.

From April to October 1968, 35 USAF attack aircraft were lost in raids in those seven months. That was not due to MiGs but non-radar-guided automatic weapons. Furthermore, AAA and AW fire was used to confuse US aircrews during attack approaches, sometimes resulting in the loss of the aircraft. When successful, this NVN method enhanced the probability of weapon delivery errors. But, the general result of this NVN tactic was a reduction in the effectiveness of weapons delivery rather than lost or damaged aircraft.

After November 1, the MiGs' ability to repair, reequip, and retrain should have posed a formidable threat to the United States forces, but it did not. The reason for that was due to the GCI and SAM system established by the US Navy, which, in conjunction with USAF and USN aircraft, provided the US aircraft with the same type of defense that US pilots encountered when they flew into the NVN cauldron.

Thus the end of the 'air war' had arrived. Following the USAF victories of 1968, there was an interruption in shootdowns that lasted three and a half years. The subsequent USAF aerial victory over a MiG would not come until February 21, 1972. Numerous changes took place during this period; the election of President Richard M. Nixon, the withdrawal of the bulk of American forces from South Vietnam, and a renewed emphasis on turning over responsibility for the conduct of the war to the South Vietnamese armed forces.

THE 1972 EASTER OFFENSIVE

The extremely long lull in offensive air activity ended on February 21, 1972, when the USAF got its first kill.

On March 30, 1972, the North Vietnamese began the 'Nguyen Hue Offensive' in honor of a Vietnamese hero who had inflicted a massive defeat on Chinese forces in 1789. It is also known as the 'Easter Offensive.'

Two other NVA military thrusts emerged from Cambodia into South Vietnam within two weeks. One presented a threat to Saigon, while the other, located in the Central Highlands, sought to split South Vietnam in two. The NVA assault ultimately comprised over 140,000 troops, 600 tanks, armored vehicles, and vastly improved mobile air defense capabilities, including new shoulder-fired surface-to-air missiles (SAMs).

When the attack began, fewer than 70,000 US military personnel were stationed in Vietnam. That was a low number compared to the peak in 1967 when the total number was 530,104.

American military advisors were instrumental in supporting retreating South Vietnamese soldiers. Moreover, after incurring tremendous losses, the Army of the Republic of Vietnam (ARVN), with the aid of US air support, delayed and then reversed the advance of the NVA by the end of July. The ARVN and US Army stabilized the front in Quang Tri Province and repelled strikes at Hue, Kontum, and Saigon.

In the offensive, the North Vietnamese lost up to 100,000 soldiers and a significant amount of armor and equipment. The approximate number of ARVN casualties was 43,000, including 10,000 killed. Furthermore, the Easter Offensive resulted in the deaths of around 25,000 Vietnamese civilians, and about one million more were left homeless.

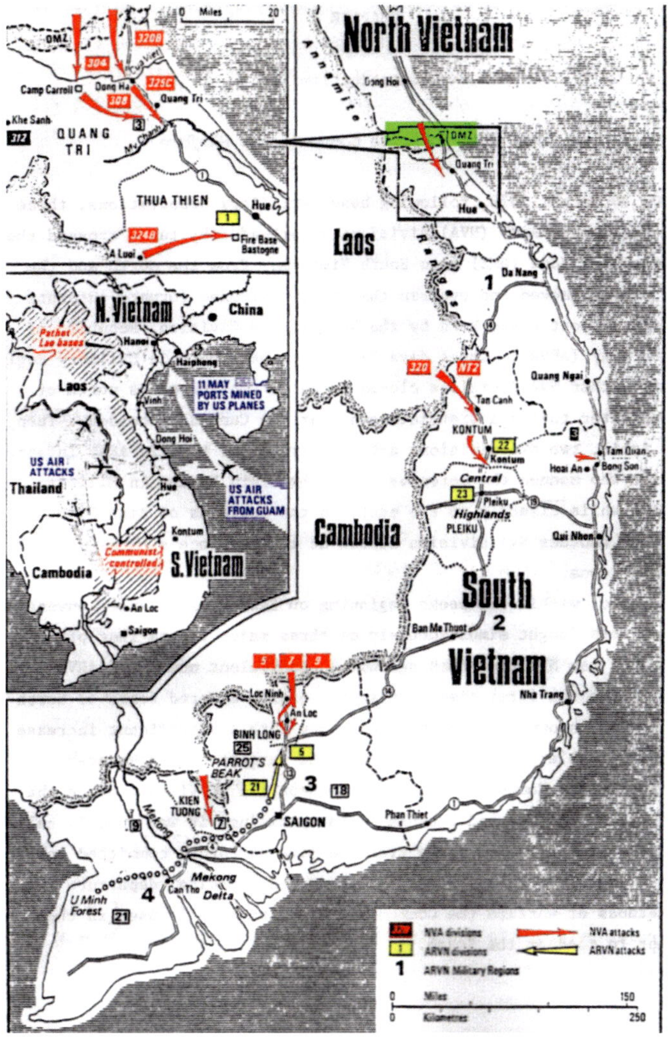

This map represents the 'Nguyen Hue Offensive' and the meticulously planned attacks of the NVA. The red rectangles show the NVA's movements, while the yellow ones portray the Army of the Republic of Vietnam's (ARVN) response. As can be seen, the ARVN was overwhelmed by the multiple attacks. (Map: US Army)

The ARVN won the 'Easter Offensive' with the help of the American forces. However, that was not completely the final result. President Nixon's administration had been holding peace negotiations in Paris with the North Vietnamese to remove both the United States and the North from South Vietnam and leave the South intact. President Nixon's agreement included a promise to return with air and naval power if the NVN did not comply.

In a later meeting, Hanoi's 'Easter Offensive' neared its peak as Kissinger met his North Vietnamese negotiating counterpart, Le Duc Tho, in Paris on May 2.

Meanwhile, in the 'Easter Offensive,' South Vietnamese forces abandoned their northernmost provincial capital, Quang Tri, the day before. The NVN Government smelt blood and was confident that victory was near. Also, according to Kissinger's memoirs, the North's delegation in Paris became intransigent and insulting. The change in attitude by the North Vietnamese incensed Nixon, who ordered the Joint Chiefs to present him with options.

OPERATION LINEBACKER I - MAY 9, 1972

In essence, 'Linebacker I' began long before it was introduced. There have been contingency plans for such an operation for years. These plans had been regularly revised to account for the shifting political atmosphere, tempo of the conflict, advancements in weaponry, and enemy action or response. As soon as the North Vietnamese invasion began and the enemy's intentions became known, the President, his National Security Council, and the Joint Chiefs of Staff prepared for an all-out campaign to repress the North Vietnamese attack and to pave the way for honest peace discussions. In a telegram dated April 4, 1972, the Joint Chiefs of Staff (JCS) referenced a new set of rules about the Vietnam War. They sought suggestions on how to have the most significant possible impact on the adversary through the creative implementation of new initiatives.

The result was an operation authorized by the President after he suspended peace talks in Paris on May 8, 1972. For the first time, the United States imposed a naval blockade and mined the waters of Haiphong and other North Vietnamese ports. Simultaneously, the President authorized a renewal of air strikes throughout North Vietnam above the 20th parallel. The operation for the air strikes was named 'Linebacker I.'

Boeing B-52D-60-BO Stratofortress (s/n 55-0100) dropping bombs over Vietnam during 'Operation Linebacker II'. (USAF Museum)

THE MiG-21 DURING LINEBACKER I

At the beginning of 1972, the NVA AF discovered that the USAF had analyzed the strengths and weaknesses of the MiG-21 and was able to develop tactical strategies for combating the Soviet-designed jet. The F-4 Phantoms, for example, sought to combat the MiGs on turns to keep the conflict as low as possible and prevent shifting the battle to a higher altitude.

As mentioned, the F-4 Phantom outperformed enemy fighters in the horizontal sphere, only at altitudes less than 15,000 feet (4,570 m). The MiG maintained its advantage at middle and high altitudes. When attacking the MiG, F-4 pilots were advised to fly as low as possible before maneuvering near the MiG's blind spot. Furthermore, the separation between two F-4 Phantoms engaged in combat with MiG-21s was raised to 2,950 ft. (900 m), allowing Phantom crews to better manage the airspace. However, the new tactics were practically untried. Were they going to be better than the last ones? The USAF and the US Navy supplied the answers to these questions.

MAY 10, 1972

The most intense day of air warfare over North Vietnam occurred on May 10, 1972. On that day, the US Navy and US Air Force shot down 11 MiGs over North Vietnam, with the US forces losing four of their own; two Navy and two Air Force F-4 Phantoms. Moreover, the American forces ensured total air superiority from that day on.

On the morning of May 10, American aircraft attacked Haiphong, Van Ninh, Pha Lai, Son Doong, and Long An, as they were devoid of air protection. Tu Ky, however, was avoided, as that city had MiG patrols protecting it. A pair of MiG-21s took off from the Kep airfield to intercept the Americans, but one was struck by AIM-7 missiles from Phantom F-4Ds of the 432nd Tactical Reconnaissance Wing's 555th Tactical Fighter Squadron as it was taking off and was destroyed. However, the second MiG-21 with NVN AF 'Ace' Dang Ngoc Nga in the cockpit avoided being hit by a missile (Dang Ngoc Nga had seven air victories in air battles during the war). First, he gained a height of 3,280 ft. (1,000 m), then he warned the pilots in the two MiG-21s taxiing for takeoff behind him of the danger.

Suddenly, the Vietnamese ace spotted a pair of Phantoms just before him. Dang Ngoc Nga dropped his fuel tank to give chase, but the tank collided with and damaged the control surfaces of his MiG during separation. Nonetheless, he did not abandon the pursuit. Instead, he increased his speed from 560 to 685 miles per hour (900 to 1,100 kph). The Phantoms became aware of the threat and dispersed. One went into a climb while the other made a horizontal turn. Nga rushed after the second one.

He launched a K-13 missile at the Phantom from a distance of 3,950 feet (1,200 meters), and it hit its target. The downed aircraft was an F-4E-36-MC from 58th TFS, 432nd TRS. It went down with both crew members who were killed. The MiG pilot felt that one downed Phantom was not enough, so he spotted the second Phantom and started giving chase again. The crew members of the second Phantom were in luck, as the second K-13 missile failed to launch owing to a launcher malfunction. The MiG veteran flew back to the Noi Bai airfield at 9:12 am.

Strikes on sites north of Hanoi were carried out by vast groups of Phantoms, attacking the airbases of Noi Bai, Hoa Lac, and Kep, where MiGs were deployed. Two MiG-21s from the 921st squadron performed an almost suicidal takeoff from the Noi Bai airstrip so that they could take on the F-4s. Six Phantoms were intercepted over Tuyen Quang in North Vietnam's northeast. One of the MiGs' wingmen shot down an F-4, but an NVN anti-aircraft missile shortly shot him down in a friendly fire blunder. Despite the Vietnamese assertion of the Phantom kill, the US did not confirm the loss of the F-4.

In another incident, the pilot of a pair of MiGs that received 16 20 mm bullets fired from an American F-4E fighter's six-barreled Vulcan cannon unexpectedly survived. The pilot managed to return to the Noi Bai Air Base.

Operation Linebacker I was a victory for the Americans. North Vietnamese MiGs battled hard throughout Linebacker I, as they did during Rolling Thunder, but they could not take control of the skies. That was partly due to American radar detecting the interceptors rising off the runways, flying in the air, or being on the ground, allowing controllers to order Air Force F-4s and Navy fighters against them.

In a change made during the SEA conflict, each aerial victory was given to the pilot and the weapon systems officer. Thus, two 'backseaters,' Captains Charles B. DeBellevue and Jeffrey S. Feinstein, but only one pilot, Capt. Richard S. 'Steve' Ritchie became Air Aces due to changes in these pairings. All three flight crews made their fifth kill during Linebacker I, which lasted until October 1972. That was when President Nixon, buoyed by progress in ceasefire negotiations, limited the bombing to the southern part of North Vietnam.

A photo taken from above Takhli Royal Thai Air Force Base during 'Operation Linebacker' in 1972. Two General Dynamics F-111As from the 429th Tactical Fighter Squadron, 474th Tactical Fighter Wing are visible in the foreground. The remaining 25 aircraft are F-4D Phantom IIs of the 49th TFW. (USAF)

Richard Nixon was re-elected President on November 7, 1972, with 520 of 538 electoral votes. The leaders in Hanoi may have believed that the new Congress, reflecting the sentiment of the American people, would decide that the United States had done enough for South Vietnam and eliminate or drastically reduce aid to the government in Saigon. That would pave the way for the North Vietnamese to seize control of the South. Nixon, however, was a political animal and a genuine master of the political game. On December 14, 1972, before Congress convened, President Nixon issued an ultimatum to the North Vietnamese. To resume meaningful talks within 72 hours or face the implications of a major bombing campaign. However, Hanoi resisted, prompting President Nixon, with the approval of the Joint Chiefs of Staff, to launch 'Operation Linebacker II' on December 18.

LINEBACKER II – THE END GAME

In terms of scale, the Linebacker II bombing campaign was unprecedented in aviation history since the bombardment of German cities by allied aircraft during World War II. Moreover, the bombing raids were prepared with extensive aerial surveillance of Vietnamese airfields, anti-aircraft missile battalions, and anti-aircraft artillery positions.

On December 18, at 7.40 pm, B-52 strategic bombers and fighter-bombers struck the Kep, Noi Bai, Gia Lam, Hoa Lc, and Yen Bai Air Bases, as well as the industrial areas of Dong An, Yen Vien, Dak Giang, and the Migri Radio Transmitting Center. At the same time, Haiphong, North Vietnam's major port, was bombarded.

Elsewhere, F-111 variable sweep fighter-bombers attacked airfields, while the F-4 Phantoms suppressed the ground air defense systems. B-52 bombers, backed by around 100 F-4 Phantoms, flew from Tamdao to Hanoi and bombarded the NVN capital.

Four F-105G Wild Weasels were used to strike the Vietnamese radars and were effective. At about 6 pm, radar installed in the Quang Binh Station reported intense interference in the frequency range. At 6.30 pm, a similar report came from the headquarters of the 45th radar company to the Zetas from the 16th radar company. Nevertheless, the radiometers managed to detect a large group of B-52 bombers. At 7.25 pm, a trio of F-111s appeared over Noi Bai Air Base and started bombing; however, they failed to destroy the runway.

A MiG-21MF took off to intercept the F-111s. The North Vietnamese pilot noticed many American aircraft heading for Hanoi. He flew toward the B-52s, whose pilots turned on their air navigation lights to avoid colliding with the MiG-21. The pilot of the MiG-21 activated his afterburners and prepared to launch his missiles. He also started the RP-21 radar sight, but his screen was cluttered with interference. He couldn't fire a missile in that situation, so he gave up and returned to base. Later that night, MiGs from the Hoa Lak Air Base also took to the air, but the pilots failed to detect the enemy.

The B-52s bombarded Hanoi and Haiphong again the next night. The Vietnamese quickly adapted to their adversary's tactics, and on December 23, NVN Air Force fighters downed four Phantoms, losing one MiG-21. The USAF claimed three additional destroyed MiG-21s in December, on December 21 and 22 (both were shot down by Phantom crews from the

This is a Wild Weasel F-4C of 67 TFSq, 18 TFWing, based at Korat on 'Temporary Duty' from Kadena September 1972 to February 1973. (USAF)

432nd Tactical Reconnaissance Wing), and on Christmas Day, December 25, a MiG was shot down by a B-52 gunner from the 307th Bombardment Wing.

By December 26, the USAF had built a decisive advantage over the North Vietnamese defenses. Sixty B-52s were deployed on each of the campaign's final three nights, December 27, 28, and 29. On December 27, two B-52s were shot down, one in North Vietnam territory, and the other in Thailand, where the crew bailed out.

On the final day of 'Operation Linebacker II', December 29, the USAF's bomber and support crews were in top condition. The adversary, on the other hand, was clearly in turmoil, able to fire just 23 SAMs at the American planes. The NVN troops were nearly out of SAMs, their MiGs were grounded, and their radar and radio lines were broken. In summary, it was the end for the NVN forces since they had now reached a dead-end street.

DECEITFUL PEACE

The peace treaty was finally signed in Paris on January 27, 1973. However, it was more of a ceasefire than a genuine settlement since it left North Vietnamese forces in possession of territory within South Vietnam, from which they might restart the war.

South Vietnam's existence depended on the attitudes of the two powers; North Vietnam's readiness to abide by the peace treaty and the United States' determination to follow President Nixon's vow to act fast and strongly if Hanoi once again attacked South Vietnam.

However, the Chief of the US Air Force, General John William Vogt Jr., doubted that the US would respond to a future invasion with the same massive aerial firepower unleashed in 1972. He claimed that General Haig, who was still on Kissinger's staff at the time, asked him to use his influence with other South Vietnamese generals to reassure them of America's commitment. He flatly refused, as he would be seen as a deceiver. General Vogt knew that the American promise would remain unfulfilled.

The Vietnam War was initially Americanized and later became Vietnamized. It concluded in January 1973 for the United States but not for the South Vietnamese. Two years later, on April 30, 1975, the North Vietnamese armed forces overran South Vietnam and took control of the whole country.

COUNTRIES	MiG VARIANTS (YEAR DELIVERED, MODEL, & APPROXIMATE NUMBER IN SERVICE)
Afghanistan	1973, 30+ MiG-21F-13. 1979, 70 MiG-21MF, 50+ MiG-21bis, 6 MiG -21UM 160+
Albania	12 Chengdu F-7A (Chinese copy of the MiG-21) 12
Algeria	1967, 30 MIG-21PF, 40+ MIG-21PFM. 1977, 6 MIG-21MF. 1978, 25 MIG-21MF 100
Angola	1976, 12-MIG-21MF, 8 MIG-21F-13, 2 MiG-21US. 1980, 12 MiG-21MF, 2 MIG-21US. 1983, 15 MiG-21bis. 1985, 30 MiG-21bis. 70
Bangladesh	1973,12 MIG-21MF, 2 MIG-21UM. 1990, 9 Chengdu 7MB, 4 Guizhou FT-7B. 2006, Chengdu F-7BG, F-7MB 37
Bulgaria	1963, 10+ MiG-21F-13 and MiG-21U / UM, (in 1974 15 F-13s converted to MiG-21F-13R), 1965, 12 MiG-21PF, & MiG-21PFM, 1969, 12+ MiG-21M, 1970? first MiG-21US, 1974-1982, MiG-21UM, 1983-1985, MiG-21bis delivered 200+
Burkina Faso	1985, 8 MiG-21bis, 2 MiG-21UM 10
Cambodia	1982, 19 MG-21bis, 3 MiG-21UM, 1996, 12 MiG-21s & 2 MiG-21UMs were converted to MiG-21-2000s by Israeli company. 22
China	1965-1969, China builds MiG-21 fighter under license, designated:'J-7', and the'JJ-7' two-seat trainer version. In 1986, more than 400 J-7s were produced in total. By 2007, China had built about 1,575 J-7s
Congo (Republic)	1986, 14 MiG-21bis and 2 MiG-21UM. 1988, 1 MiG-21bis 17
Croatia	1995, Croatia bought 40+ MiG-21bis & UM s from the Ukraine. In 2003, 8 MiG-21 bis & 4 MiG-21UMs were upgraded in Romania 40+
Cuba	Post Cuban Missile Crisis, Cuba received 46 MiG-21F-13 & 2 MiG-21UM. 1966-1967, 30 MiG-21 PFMs, which were replaced by MiG-21 MFs in 1968, with a further 12 MiG-21Rs arriving in the same year. 1972-1974, 60 MiG- 21MFs. From the 'F-13' onwards, Cuba received all main variants of the MiG-21, making it the first country in the Americas to do so. 150+
Czechoslovakia	Czechoslovakia was licensed to build the MiG-21. From 1963 -1973, Aero Vodochdy constructed 194 MiG 21 F 13s and branded them as Aero S-106. Then, between 1981-1983, it produced another 220 MiG-21MFs, MiG-21Fs , MiG-21Us, and a batch of 80 MiG-21Rs. 350+
Egypt	1962, First delivery of MiG-21F-13s, 1964, Egypt received MiG-21 PFs. 1967, delivery of 50 MiG-21PFs, PFLs and PFMs which came together with R-3S infrared homing missiles. Egypt had roughly 230 MiG-21s at the commencement of the Six-Day War in 1967, 75 of which were MiG-21 PFMs delivered in 1966. Over a hundred of these fighters were destroyed during the conflict, most of them on the ground, without firing a shot. To make up for these losses, Czechoslovakia provided some MiG-21 F-13s for Egypt. Moreover, from 1970-1972, the USSR provided Egypt with additional MiG-21s. In the Yom Kippur War of October 6,1973, the fighting was slightly more even. At the end of that conflict, in aerial action, the Egyptian air force lost 73 MiG-21s and destroyed 27 Israeli aircraft. When Soviet assistance to Egypt ended in 1972, so did deliveries of the MiG-21s, which were replaced by 80 Chinese Chengdu F-7Bs in 1980. Egypt had three MiG-21 squadrons and two F-7B squadrons up to 2007. Total MiG-21s: 475, J-7s: 80
Ethiopia	1977, 48 MiG-21MFs delivered. 1982-83, deliveries of 100 more types which included MiG-21MFs, UMs, and some MiG-21Rs. 1986-88, 30 MiG-21bis received. 100+
Finland	Finland was the first non-Warsaw Pact state to receive MiG-21s. 1963, 21 MiG-21F-13s delivered. 1965, 2 MiG-21Us arrive. 1971-1973, 9 MiG-21 F-13s converted to reconnaissance variant. 1974, several MiG-21UMs delivered. From 1978, the F-13s were being replaced by the MiG-21bis. A total of 26 MiG-21'bis' fighters were delivered. Six MiG-21bis were upgraded into the recon variant and designated MiG-21bisT. 1981, delivery of two-seater MiG-21UM trainers. 1998, the last MiG-21bis was withdrawn from service. 60
German Democratic Republic	The German Democratic Republic was a major user of the MiG-21 series. April 1962, first deliveries of 76 MiG-21F-13s to the following regiments: JG-8 (Neuhardenberg), JG-9 (Peenemünde), and the JG-3 (Nays-Malksetalia). The last MiG-21F-13s were 'withdrawn from use' (WFU) in 1985. In the spring of 1964, 54 MiG-21PFs were delivered to JG1 & JG-8 regiments. Some of the PFs were deployed to JG 2 and 9 'Heinrich Rau (Peenemünde). The remaining MiG-21 PFs were retired from service in 1988, while the PFMs were wfu in 1990. From April 1965 to July 1967, 45 MiG-21U 'Mongols' were

	delivered to FAG-15 Training Unit 'Heinz Kapelle' (Rothenburg-Oberlausitz). Apart from JG-8, the following regiments received the MiG-21U: JG1, JG3, JG7, and JG9. 1968, 134 MiG-21PFMs delivered. 1968-1970, 17 MiG-21US trainers delivered. Also, the MiG-21UM was delivered 1970-1971, and 3 further aircraft were delivered in 1978 to complete the order. In late fall 1968, 87 MiG-21Ms, the export version of the MiG-21SM, delivered to JG-8 regiment, and other units. In 1973, the GDR's MiG-21M fleet was reduced by the transfer of 12 jets to Syria, to compensate for losses sustained during the many air battles fought against Israeli aircraft. Between 1970-1971, more MiG-21MFs were delivered. The first batch of 14 were delivered to the JG-3 regiment, while another 36 were delivered to other regiments. There was a final batch of 12 more MiG-21MFs in 1975. From October 1975 to May 1978, The GDR AF received 46 MiG-21bis fighters. These were the final MiG-21 deliveries to the East German Air Force. 450+
Guinea	1986, delivery of 8 MiG-21MFs which formed one combat unit. 8
Guinea-Bissau	Delivery of several MiG-21PFMs. Delivery date unknown, probably in the 1980s. Possibly 6 units
Hungary	1961, delivery of 12 MiG-21F-13s, followed by another 68 units, totaling 80 aircraft. 1963, 24 MiG-21PFs received. Also, the first of 16 MiG-21U trainers were delivered in 1963. Furthermore, 24 units of the two-seat variant MiG-21UM were delivered in stages until 1986. 1971, delivery of 48 MiG-21MFs. 1975, received first batch of MiG-21bis, out of a total of 62 units. 200+
India	India was the other country, apart from China and Czechoslovakia, which had the license to produce the MiG-21. 1963, first batch of 10 MiG-21F-13s arrive. In the fall of 1964, 4 more MiG-21F-13s and 2 MiG-21PFs delivered. They were just in time for the India-Pakistan Kashmir War of 1965, but they did not play a major part, due to lack of experienced Indian MiG-21 pilots, at the time. During the war, two MiG-21PFs delivered to India at the end of 1964 were destroyed on the ground by Pakistani F-86 Sabres on September 6, 1965. 1966-1974: Preparation and manufacturing by India of the MiG-21FL under license at the HAL Corporation's Nasik factories with a total production of 200 units, as per the contract. Prior to 1964, MiGs were assembled from components provided by the USSR. Then, in 1969, independent production began with an annual output of around 30 units. During the Indo-Pakistani War in December 1971, the Indian MiG-21 pilots were adequately prepared and performed their duties against the Pakistani and Jordanian Lockheed F-104s, by shooting down three enemy aircraft between December 12 and 17. 1973-1979, A new contract for the licensed manufacturing of the MiG-21M was started in 1973 and ran from 1973-1979. Up until 1975, just 10 planes were produced annually. However, with the addition of the USSR's supplies, a total of 150 MiG-21Ms were built by 1979. Additionally, work on MiG-21 engine maintenance and assembly was conducted at the Koraput plant. February 14, 1973, saw the first flight of the Indian-assembled MiG-21M. By 1991, the total number of MiGs Produced by India, totaled 675.
Indonesia	1962, delivery of just over 20 MiG-21F-13s. 20+
Iran	After severing ties with the United States in 1979 and not having a positive relationship with the Soviet Union, Iran turned to China in 1985 and ordered 18 Chengdu F-7Ms (license-built version of the MiG-21), and 4 Guizhou FT-7s (the third generation export version of the F-7). The aircraft were delivered in 1987. In 1993, a further 25 F-7Ns and FT-7s were bought. During the Gulf War of 1991, Iran obtained several MiG-21s that were among the hundreds of Iraqi military and civil aircraft that sought refuge in Iran during the war, and which Iranian authorities refused to return to Iraq. 60+
Iraq	Iraq was a member of the non-aligned states, which favored the USSR more than the USA. Thus, in 1963, it received 35 MiG-21F-13s. However, one of the MiG-21s was flown to Israel by an Iraqi defector, Cpt. Munir Redfa, on 16 August 1966. The Israelis struck gold with the receival of the MiG-21, one year before the Six Day War. The Israeli AF got to know the Soviet whizz jet, inside-out, before that conflict. 1966, Iraq purchased around 30 MiG-21PFs. In 1970 it made the biggest purchase ever, with an order of 55 MiG-21PFMs. However, several PFMs were redirected to the Egyptian, and Syrian Air Forces to replace their destroyed MiGs from the Six Day war. Then, in the 1973 Yom Kippur War, another 18 MiG-21PFMs were delivered to the Syrian AF. During that war, Iraq received another 40 MiG-21PFMs, many of which were diverted to the Egyptian and Syrian Air Forces. Finally, in the 1980-1988 Iran-Iraq War, Iraq used what remained of its MiG-21PFMs against Iran. 1979, Iraq received 15 MiG-21Rs for recon duties over Iran. 1982, delivery of 70 F-7Bs for the Iraqi AF from China. 1983-1984, 61 MiG-21bis delivered. They were the last MiGs it purchased, due to the embargo placed on Iraq after Operation Desert Storm 1990-1991. 310+

Israel	With the exception of the defector MiG-21F-13 in 1966. During the 1967 Six Day War, a group of six Algerian Air Force pilots, flying MiG-21F-13s made a mistake and landed in an Israeli controlled region in Sinai, and their MiGs were seized by the Israeli Air Force. 7
Korea (North)	North Korea's stats are difficult to establish due to the country's lack of transparency. In 1963, 14 MiG-21F-13s were in service. 1967, North Korea acquired 80 MiG-21F-13s. In 1970, the North Korean Air Force (DPRK AF) had 96 MiG variants, 46 of which were 'PF' and 'PFM' versions. By 1975, just 24 MiG-21s remained in service. More deliveries from the USSR were then made, and by 1983, 120 different MiG-21s were in service, with further supplies reaching 200 aircraft in 1986. The number of MiG-21s in service peaked at 220 in 1991, when the DPRK AF received 30 Chengdu F-7Bs from China. In 1999, the DPRK AF acquired 38 more ex-Kazakhstan AF MiG-21bis. 220+
Laos	In the early 1970s, Laos received about 30 MiG-21Fs from the Soviet Union, and some MiG-21U two seater trainers. In 1991, it had a total of 44 MiG-21s, which were in operation with two fighter squadrons based at Wattay Airport, near the capital city, Vientiane. 44
Libya	1975, 50 MiG-21MFs and a few MiG-21UMs were delivered to Libya. It dispatched 30 MiG-21MFs to Syria to restock after suffering losses in the 1973 Yom Kippur War. Egypt and Libya had a border dispute in 1977, when the two countries traded blows. The Egyptian-Libyan War, commonly known as the 'Four Day War,' took place between July 21 and 24, pitting Egyptian M-21MFs against Libyan Mig-21MFs. A ceasefire was called, and the conflict came to an uneasy end. In 1991, Libya had an all-time high total of 104 MiG-21s in service. 104
Madagascar	In 1980, the first 8 MiG-21Fs out of 15 ordered were delivered to the Madagascar AF. The second batch of 7 were delivered later. 15
Mali	1974, the West African nation received 12 MiG-21bis from the USSR. 2005, the Czech Republic supplied 2 MiG-21MFs and 1 MiG-21UM trainer. 15
Mongolia	1977, delivery of the first 8 MiG-21PFs and 4 MiG-21UM trainers. This was the most Mig-21s in operation with the Mongolian Air Force. By 2011, 10 MiG-21s remained in service. 12
Mozambique	1978, there were 30 MiG-21MFs in service. 1982,48 Mig-21bis variants arrived in Mozambique. By 1991, a total of 43 MiG-21 variants were still in service. 70+
Myanmar	Because of its close relationship with China, Myanmar acquired solely Chinese MiG-21 models. 10 Chengdu F-7s were delivered in 1991. Then, in 1993, a further 10 were dispatched. 20+
Nigeria	1975-1976 25 MiG-21MFs and 6 MiG-21UMs were delivered. The Nigerian Air Force had 31 MiG-21s in service in 1991. In 1993, 22 MiG-21s of various types were in service. However, due to the end of Soviet technical assistance, all aircraft operations terminated in the early 1990s. 31
Pakistan	Pakistan, purchased many Chinese made copies of the MiG-21. 1988, delivery of 20 Chengdu F-7Ps. 1989, 15 Guizhou FT-7P delivered. 1990, 40 F-7s and 36 FT-7s received, and in service with the Pakistan AF. 100+
Poland	1961, the first deliveries of MiG-21F-13s. They were put into service one year later, in 1962. 1963, delivery of all 25 MiG-21F-13s was complete. All F-13s were wfu in 1973. The final 12 operable F-13s were supplied to Syria in 1974 to compensate for losses sustained during the Yom Kippur War. 1964-1965, delivery of 84 MiG-21PFs. 1965-1966, 11 MiG-21U trainers delivered. 1966-1968, 132 MiG-21PFM received by the Polish AF. Another 30 were delivered later, a total amount of 162 'PFM' variants. 1968-1972, the next variant purchased was the MiG-21R 'reconnaissance' with 36 aircraft being delivered. 1969-1970, saw the delivery of 12 MiG-21US training aircraft. 1971-1981, during these years, 54 MiG-21UMs were delivered. Also, from 1972 to 1975 120 MiG-21MFs were received. Then, in December 1972, all Polish fighter squadrons, previously equipped with MiG-19s were now equipped with the new MiG-21MF. 1979, saw the start of deliveries of 72 MiG-21bis fighters. After 1989, there were many aircraft withdrawn from use: 1989, MiG-21PF and MiG-21PFM, 1990, MiG-21U, 1997, MiG-21R (recon), 1999, MiG-21bis, 2002, MiG-21M, and 2003 the MiG-21US, MiG-21UM, as well as the MiG-21MF. 580 +
Romania	1963, Romanian AF receives 24 MiG-21F13s. 1965, delivery of 40 MiG-21PFs. Once in Romania, the 'PF' designation was replaced with 'RFM'. Thus, they became MiG-21RFMs. 1965, the Air Force received 4 MiG-21U trainers and 3 more arrived later. 1969, delivery of 14 MiG-21US training jets. Then between 1972 and 1980 it took delivery of 31 more trainers, the MiG-21UMs. In 1966, Romania took delivery of 29 potent MiG-21PFs. They were also re-designated as MiG-21RFMMs.This variant was nuclear capable. 1968, delivery of 11 MiG-21Rs (Romanian designation: MiG-21C). 1969, a delivery of 60 MiG-21Ms was made. 1993, the Romanian AF operated 218 MiG-21 (various fighter variants) and 10 MiG-21C (recon version). Romania signed a contract with the Israeli firm 'Elbit' for the upgrading of 100 of its MiG-21s. In 1997, the Romanian AF received its 100 fully refurbished MiG-21-2000s, at a total cost of $330 million USD. The 'new' Migs were now designated 'Lancer' 'A' for the

	fighter bombers, and Lancer 'C' for the interceptors. As for the trainer version, the MiG-21UMs purchased in the 1970s were also upgraded by Elbit and were designated Lancer 'B'. 218
Slovakia	After Czechoslovakia was divided into the Czech and Slovakia republics in 1993, Slovakia acquired 70 former Czechoslovakian MiG-21s of various types. But, at the beginning of the new millennium, it was left with 8 MiG-21MFs. In January 2001, all were withdrawn from use. 70
Somalia	1974, delivery of 10 MiG-21MFs, and 4 MiG-21UM two-seaters. Later, due to a war with Ethiopia, and, an ensuing civil war, there were no other MiG deliveries. 14
Sri Lanka	Another customer of the Chinese MiGs was Sri Lanka. In 1991 it received 4 Chengdu F-7Bs, and 2 Ghuizhou FT-7B two seater trainers. 2008, Sri Lanka receives another 6 F-7Gs. In 2022, the Air Force still had 5 F-7s in service. 12
Sudan	1971, 18 MiG-21MFs delivered, together with 4 MiG-21U trainers. In 2022 Sudan had 4 MiG-21s, and 20 Chinese F-7s in service (most probably the highly rated F-7M version). 40+
Syria	The pro-Soviet, and now pro-Russian stalwart Syria, has had many MiGs in its fleet. 1961, delivery of 36 MiG-21F-13s, followed by 45 more in 1965. It later received 15 MiG-21FLs and 6 Mig-21Us in 1966. After the Six Day War, the Soviet Union supplied a further 24 MiG-21s to replenish the Syrian AF. 1968, Syria received 100 MiG-21PFMs. 1970, delivery of 6 MiG-21Rs. In 1971, 21 MiG-21MFs were delivered, with another 60 of the type in 1973, as well as 20 MiG-21UMs. After the end of the October war in 1973, the USSR once more sent 75 MiG-21 MFs to the Syrian AF, to make up for its losses. On June 10, 1982, in skirmishes above Lebanon, the Syrian AF lost 10 MiG-21bis. The Syrians received more assistance from the Soviet Union in the 1980s in the form of 198 MiG-21bis. 600+
Tanzania	1974, the Soviet Union sent Tanzania 14 MiG-21MFs, and 2 MiG-21Us. However, in 1980, the Tanzanians turned to China for their next batch of jet fighters. They purchased 10 Chengdu F-7Bs and 2 Guizhou FT-7s. Later, in 1997, Tanzania took delivery of 4 MiG-21 MFs from the Ukrainian AF. In 2022, the Tanzanian AF still had 11 F-7s in the fleet.
Ukraine	After the collapse of the Soviet Union in 1991, the Ukraine received several MiG-21s from air bases vacated by the Russians. None of the MiG-21s served in the Ukrainian AF for too long. They were sold off to South East Asian and African countries.
USA	The United States, received some confiscated MiG-21s in 1967. Six were seized by the Israelis in the Sinai, during the Six Day War. (See the Israel information above)
USSR	The total production of Soviet built MiG-21s for Soviet use only, is immense. The total MiG-21 production in the famous Gorky Plant in Moscow was: 5,278 units, for the 'Znamya Truda' Plant (Moscow Aircraft Production Association), also in Moscow, it was: 3,203 units, and for the Tbilisi Plant in Georgia, it was: 1,677 units. Giving a total of: 10,158 MiG-21s for use by the Soviet Forces.
Vietnam (North)	The North Vietnamese AF received its first batch of MiG-21F-13 in late 1965. Two squadrons were formed in the 921st IAP. The fighter made its debut in the war in 1966. 1966, USSR delivers 30 MiG-21PFLs (Vietnamese variant). Soon after, a batch of 100 MiG-21PFMs were delivered. 1970, 60 MiG-21MFs delivered. 200+
Yemen (North)	1979-1980, delivery of 45 MiG-21 MFs and MiG-21UM training jets.
Yemen (South)	1971, first batch of MiG-21F-13s delivered, with more deliveries made later.
Yemen Republic	After their war ended on May 22, 1990, the two Yemens united. The two former enemies combined their MiG-21s, totaling more than 70 MiGs, the majority of which were the MiG-21MF variant. 70+
Yugoslavia	1962, the first batch of 45 MiG-21F-13, designated L-12s, was delivered. (the 'L—' designation was the singe seater, and the 'LL—' designation was for the two seater trainers. 1965, received 9 MiG-21Us (NL-12). 1968, delivery of 36 MiG-21PMFs (L-13). 1969, received 9 MiG-21USs (NL-14). 1970, Yugoslavia received 12 units of the first recon model, MiG-21R (L-14I). Additionally, it received 25 MiG-21Ms (L-15) that year. The next supply in 1975 was of 6 MiG-21MFs (L-15M). 1977, the start of delivery of the MiG-21bis (L-17), and the MiG-21UMs (NL-16). The last two batches were the final MiG-21 deliveries made to Yugoslavia. 200
Zambia	1976, received 16 MiG-21MFs and 2 MiG-21U trainers. 18
Zimbabwe	1986, received 12 Chengdu F-7IINs (Zimbabwe variant). 1991, delivery of 2 trainer Guizhou FT-7BZs. An FT-7Z converted in Zimbabwe in 1991 with reworked rear fuselage to withstand the higher temperature of the engine. In 2022, Zimbabwe had 7 F-7s in its fleet, possibly F-7IINs. 14

PHOTO CREDITS

Front Cover: "MiG-21MF Holzdorf" flickr photo by **Rob Schleiffert** https://flickr.com/photos/109661044@N07/22822197646 shared under a Creative Commons (BY-SA) license 2.0. Brightened ,added tone.

Page 8 **bagera3005** (https://www.deviantart.com/bagera3005/art/Mikoyan-Gurevich-Ye-2a-204582440) darkened. https://creativecommons.org/licenses/by-sa/3.0/legalcode

Page 22. **Newresid** stephanelhernault@yahoo.fr (https://commons.wikimedia.org/wiki/File:Mig21F13web.jpg), „Mig21F13web", sharpened, enhanced color, darkened by Theodore Gianna, https://creativecommons.org/licenses/by-sa/3.0/legalcode

Page 24. **bagera3005** (https://www.deviantart.com/bagera3005/art/MikoyanGurevich-MiG-21F-IRIAF-159014346) darkened, sharpened. https://creativecommons.org/licenses/by-sa/3.0/legalcode

Page 29. **Varga Attila** (https://commons.wikimedia.org/wiki/File:Tumansky_R-11.jpg), „Tumansky R-11", https://creativecommons.org/licenses/by-sa/3.0/legalcode

Page 30. "MiG-21F-13 East Germany" flickr photo by **Rob Schleiffert** https://flickr.com/photos/109661044@N07/16837045032 shared under a Creative Commons (BY-SA) license 2.0, Enhanced color, brightness, tonal width, and sharpness, by Theodore Gianna

Page 30. "Mikoyan MiG-21F-13 '2015'" flickr photo by **HawkeyeUK** https://flickr.com/photos/ajw1970/13450474033 shared under a Creative Commons (BY-SA) license 2.0, Enhanced color, brightness, tonal width, and sharpness, by Theodore Gianna

Page 31. "MiG-21F-13 Czech AF" flickr photo by **Rob Schleiffert** https://flickr.com/photos/109661044@N07/16218212883 shared under a Creative Commons (BY-SA) license 2.0, Enhanced color, brightness, tonal width, and sharpness by Theodore Gianna,

Page 31."Mikoyan MiG-21F-13 '2015'" flickr photo by **HawkeyeUK** https://flickr.com/photos/ajw1970/13450474033 shared under a Creative Commons (BY-SA) license 2.0, enhanced brightness, color, and tonal width.

Page 34. **Museum Pusat TNI AU Dirgantara Mandala Yogyakarta** (https://commons.wikimedia.org/wiki/File:MiG-21F-13-Fishbed-C.jpg), Enhanced color, brightness, tonal width, and sharpness. by Theodore Gianna, https://creativecommons.org/licenses/by-sa/4.0/legalcode

Page 34. "Mikoyan-Gurevich MiG-21F-13 '1101'" flickr photo by **HawkeyeUK** https://flickr.com/photos/ajw1970/26430929504 shared under a Creative Commons (BY-SA) license 2.0, Enhanced color, brightness, tonal width, and sharpness, by Theodore Gianna.

Page 39. "Mikoyan MiG-21PF Fishbed-D '0615'" flickr photo by **HawkeyeUK** https://flickr.com/photos/ajw1970/11016054715 shared under a Creative Commons (BY-SA) license 2.0. Enhanced color, brightness, tonal width, and sharpness, by Theodore Gianna

Page 39. Newresid **stephanelhernault@yahoo.fr** (https://commons.wikimedia.org/wiki/File:Mig21PFfamilyweb.jpg), „Mig21PFfamilyweb", enhanced color, brightened by Theodore Gianna, https://creativecommons.org/licenses/by-sa/3.0/legalcode

Page 39. "Mikoyan-Gurevich MiG-21PF Fishbed D" flickr photo by **mikelachance816** https://flickr.com/photos/mjl816/46676760404 shared under a Creative Commons (BY) license 2.0. Enhanced color, brightness, tonal width, and sharpness, by Theodore Gianna

Page 38. "Mikoyan MiG-21PF Fishbed-D '0302'" flickr photo by **HawkeyeUK** https://flickr.com/photos/ajw1970/8170584194 shared under a Creative Commons (BY-SA) license 2.0. Enhanced color, brightness, tonal width, and sharpness, by Theodore Gianna

Page 41. "Mikoyan MiG-21PF '1702'" flickr photo by **HawkeyeUK** https://flickr.com/photos/ajw1970/13450747894 shared under a Creative Commons (BY-SA) license 2.0 . Enhanced color, brightness, tonal width, and sharpness, by Theodore Gianna

Page 42. "Mig aircraft at Vietnam Military History Museum Hanoi" flickr photo by **dronepicr** https://flickr.com/photos/132646954@N02/25672857528 shared under a Creative Commons (BY) license 2.0. Enhanced color, high level of brightness, tonal width, and sharpness, by Theodore Gianna

Page 46. **aeroprints.com** (https://commons.wikimedia.org/wiki/File:C499_Mikoyan_Mig-21FL_Indian_Air_Force_(8448372348).jpg), „C499 Mikoyan Mig-21FL Indian Air Force (8448372348)", Enhanced color, brightness, tonal width, and sharpness by Theodore Gianna, https://creativecommons.org/licenses/by-sa/3.0/legalcode 3.0.

Page 49. **Jim van de Burgt NL** https://tinyurl.com/594xnr8a

Page 50. **LeCaire** https://commons.wikimedia.org/wiki/File:Egyptian_MiG-21.jpg (Public Domain photo)

Page 50. **Wistula** (https://commons.wikimedia.org/wiki/File:Soch24DSC_1027.JPG), „Soch24DSC 1027", Enhanced color, brightness, tonal width, and sharpness by Theodore Gianna, https://creativecommons.org/licenses/by/3.0/legalcode

Page 51. **Alan Wilson** from Stilton, Peterborough, Cambs, UK (https://commons.wikimedia.org/wiki/File:Mikoyan-Gurevich_MiG-21R_'1125'_(19366192611).jpg), „Mikoyan-Gurevich MiG-21R '1125' (19366192611)", Cropped, enhanced color, brightness, tonal width, and sharpness by Theodore Gianna, https://creativecommons.org/licenses/by-sa/2.0/legalcode

Page 54. **Alan Wilson** creator QS:P170,Q33132025 (https://commons.wikimedia.org/wiki/File:Mikoyan_MiG-21SM_Fishbed-J_70_red_(8460002863).jpg), „Mikoyan MiG-21SM Fishbed-J 70 red (8460002863)", Enhanced color, brightness, tonal width, and sharpness by Theodore Gianna, https://creativecommons.org/licenses/by-sa/2.0/legalcode

Page 55. **Newresid stephanelhernault@yahoo.fr** (https://commons.wikimedia.org/wiki/File:Mig21fishbedHfamilyweb.jpg), „Mig21fishbedHfamilyweb", Enhanced color, brightened by Theodore Gianna, https://creativecommons.org/licenses/by-sa/3.0/legalcode

Page 63. "MiG-21MF Preschen" flickr photo by **Rob Schleiffert** https://flickr.com/photos/109661044@N07/22398667499 shared under a Creative Commons (BY-SA) license 2.0. Enhanced color, brightness, tonal width, and sharpness by Theodore Gianna

Page 64. "FAIRFORD 26 JULY 1998 CZECH AIR FORCE MIKOYAN MIG-21MF 7701" flickr photo by **simonbutler2** https://flickr.com/photos/antonov22/50537970761 shared under a Creative Commons (BY) license 2.0. Enhanced color, brightness, tonal width, and sharpness by Theodore Gianna

Page 65. "MiG-21SMT Soviet air force hangared Altenburg" flickr photo by **Jerry Gunner** https://flickr.com/photos/13722921@N06/3117237967 shared under a Creative Commons (BY) license 2.0 Enhanced color, brightness, tonal width, and sharpness by Theodore Gianna

Page 65. "Mikoyan-Gurevich MiG-21SMT '71 blue'" flickr photo by **HawkeyeUK** https://flickr.com/photos/ajw1970/38106055645 shared under a Creative Commons (BY-SA) license 2.0. Enhanced color, brightness, tonal width, and sharpness by Theodore Gianna

Page 66. **Newresid stephanelhernault@yahoo.fr** (https://commons.wikimedia.org/wiki/File:Mig21bisweb.jpg), „Mig21bisweb", Enhanced color, brightened by Theodore Gianna, https://creativecommons.org/licenses/by-sa/3.0/legalcode

Page 66. **User:WikiABG** (https://commons.wikimedia.org/wiki/File:MiG_21_SMT_-_Altenburg-Nobitz.jpg), „MiG 21 SMT - Altenburg-Nobitz", Enhanced color, brightness, tonal width, and sharpness by Theodore Gianna, https://creativecommons.org/licenses/by-sa/2.5/legalcode

Page 70. "Mikoyan-Gurevich MiG-21bis '9204'" flickr photo by **HawkeyeUK** https://flickr.com/photos/ajw1970/18778261594 shared under a Creative Commons (BY-SA) license 2.0. Enhanced color, brightness, tonal width, and sharpness.

Page 70. "Mikoyan MiG-21bis Fishbed-L/N '48 blue'" flickr photo by **HawkeyeUK** https://flickr.com/photos/ajw1970/10041307004 shared under a Creative Commons (BY-SA) license 2.0. Enhanced color, brightness, tonal width, and sharpness.

Page 73. "Mikoyan MiG-21U-600 '2720'" flickr photo by **HawkeyeUK** https://flickr.com/photos/ajw1970/13455814034 shared under a Creative Commons (BY-SA) license 2.0. Enhanced color, brightness, tonal width, and sharpness.

Page 74. **Alan Wilson** from Stilton, Peterborough, Cambs, UK (https://commons.wikimedia.org/wiki/File:Mikoyan-Gurevich_MiG-21US_'4401'_(18716425513).jpg), „Mikoyan-Gurevich MiG-21US '4401' (18716425513)", Enhanced color, brightness, tonal width, and sharpness by Theodore Gianna, https://creativecommons.org/licenses/by-sa/2.0/legalcode

Page 75. "5141 Mikoyan MiG-21US ex Polish Air Force" flickr photo by **Paul Nelhams** https://flickr.com/photos/58611670@N04/5374771839 shared under a Creative Commons (BY-SA) license 2.0 Enhanced color, brightness, tonal width, and sharpness

Page 76. **Jerry Gunner** from Lincoln, UK (https://commons.wikimedia.org/wiki/File:9351_a_MiG-21UM_Mongol_Polish_Air_Force_Poznan-Krzesiny_(3118862346).jpg), „9351 a MiG-21UM Mongol Polish Air Force Poznan-Krzesiny (3118862346)", Enhanced color, brightness, tonal width, and sharpness. by Theodore Gianna, https://creativecommons.org/licenses/by/2.0/legalcode

Page 76. "Mikoyan-Gurevich MiG-21UM '21 red'" flickr photo by **HawkeyeUK** https://flickr.com/photos/ajw1970/38123408991 shared under a Creative Commons (BY-SA) license 2.0. Enhanced color, brightness, tonal width, and sharpness

Page 78. "2021-06-06_14-53-39.dp2.airzoo.x3f.tif.hdr.pano.2.hyper" flickr photo by **aaronx** https://flickr.com/photos/mugland/52463123541 shared under a Creative Commons (BY) license 2.0.

Page 81. **Michael Barera** (https://commons.wikimedia.org/wiki/File:Barksdale_Global_Power_Museum_September_2015_42_(Mikoyan-Gurevich_MiG-21F).jpg), Enhanced color, brightness, tonal width, and sharpness. by Theodore Gianna, https://creativecommons.org/licenses/by-sa/4.0/legalcode Enhanced color, brightness, tonal width, and sharpness. by Theodore Gianna

Page 82. "Mikoyan-Gurevich MiG-21PF "Fishbed-D"" flickr photo by **mark6mauno** https://flickr.com/photos/mark6mauno/11378872073 shared under a Creative Commons (BY) license 2.0. Enhanced color, brightness, tonal width, and sharpness. by Theodore Gianna

SOURCES

The books listed below were utilized as sources. In addition, the author conducted extensive research on the MiG-21 series in old historical aviation periodicals, journals, and newspapers and comprehensively studied USAF, USN, CIA, and other US Government reports, records, journals, and publications, all of which are too numerous to list.

Modern Fighting Aircraft, MiGs – Bill Sweetman (Salamander)

Mikoyan MiG-21 ("Fishbed") Variants – John F. Brindley (Aircraft Profile)

Mikoyan MiG-21 – Bill Gunston (Osprey)

MiG-21 'Fishbed' Yefim Gordon and Bill Gunston (Aerofax)

MiG-21 Fishbed in Action – Don Ling and Don Spering (Squadron)

Russian Aircraft Since 1940 – Jean Alexander (Putnam)

Soviet Aircraft Directory – William Green and Gordon Swanborough (The Observer's)

MiG-21, Cold War Warrior – Hans-Heiri Stapfer (Arms and Armour Press)

Jane's All The World's Aircraft (Various editions)

Made in the USA
Monee, IL
28 March 2025

14798925R00064